Die Rückkehr der „Großen Männer"
Bringing Personality Back in

Prinz-Albert-Studien
Prince Albert Studies

Herausgegeben von / edited by
Dieter J. Weiß

Band 28 / Volume 28

De Gruyter

Die Rückkehr der „Großen Männer“

Staatsmänner im Krieg – Ein deutsch-britischer Vergleich

Bringing Personality Back in:

Leadership and War – A British-German Comparison

1740 – 1945

Herausgegeben von / edited by

Brendan Simms und Karina Urbach

De Gruyter

Gedruckt mit Unterstützung von:
Niederfüllbacher Stiftung
Stadt Coburg
Sparkasse Coburg-Lichtenfels
Universität Bayreuth
Universitätsverein Bayreuth

ISBN 978-3-11-023294-3
e-ISBN 978-3-11-023295-0

Library of Congress Cataloging-in-Publication Data

Die Rückkehr der "grossen Männer" : Staatsmänner im Krieg : ein deutsch-britischer Vergleich = Bringing personality back in : leadership and war : a British-German comparison, 1740-1945 / herausgegeben von/edited by Brendan Simms und Karina Urbach.
p. cm. -- (Prinz-Albert-Studien = Prince Albert studies ; Bd. 28/v. 28)
Includes bibliographical references.
ISBN 978-3-11-023294-3 -- ISBN 978-3-11-023295-0 (ebook)
1. Great Britain--History, Military. 2. Germany--History, Military. 3. Statesmen--Great Britain--Biography. 4. Statesmen--Germany--Biography. 5. Command of troops--Cross-cultural studies. 6. Political leadership--Cross-cultural studies. 7. Leadership--Great Britain--History. 8. Leadership--Germany--History. I. Simms, Brendan. II. Urbach, Karina.
DA65.R83 2010
355.3'3041--dc22

2010025250

Bibliografische Information der Deutschen Nationalbibliothek

Die Deutsche Nationalbibliothek verzeichnet diese Publikation in der Deutschen Nationalbibliografie; detaillierte bibliografische Daten sind im Internet über http://dnb.d-nb.de abrufbar.

Satz: PTP-Berlin GmbH
Druck: Hubert & Co. GmbH & Co. KG, Göttingen
∞ Gedruckt auf säurefreiem Papier

Printed in Germany

www.degruyter.com

For Professor Franz Bosbach
President of the Prince Albert Society
from 1995 to 2008

Vorwort

Die 28. Konferenz der Prinz-Albert-Gesellschaft im Jahr 2009, die am 11. und 12. September im Riesensaal von Schloß Ehrenburg in Coburg stattfand, stand unter dem prägnant-programmatischen Titel „Die Rückkehr der großen Männer“. Dahinter steht natürlich die Anspielung auf die berühmt-berüchtigte Sentenz Heinrich von Treitschkes (1834–1896) von den Männern, die Geschichte machen.[1] Strukturen allein können den Fortgang der Geschichte nicht bestimmen, wenn man auch den Anteil des Individuellen an der historischen Entwicklung nicht generalisieren kann. Die Rückkehr der großen Männer, das deutet auch auf das in den letzten Jahren neu erwachte Verständnis für die historische Gattung der Biographie hin, die seit den siebziger Jahren des vergangenen Jahrhunderts in Verruf geraten war. Ihre Berechtigung war besonders von sozialwissenschaftlicher Seite in Zweifel gezogen worden. Dem steht das anhaltende öffentliche Interesse an Lebensbeschreibungen bedeutender Damen und Herren der Geschichte gegenüber, das – wenn Historiker von Profession dem nicht nachkommen – eben von Sachbuchautoren erfüllt wird.

Heinz Gollwitzer formulierte prägnant den Anspruch, der an eine wissenschaftliche Biographie zu stellen ist, und zeigte auf, was sie für unsere historische Erkenntnis zu leisten vermag:

> *„Die Biographie ermöglicht schließlich wie keine andere historische Gattung, Strukturen, Prozesse und Kontingenz zueinander in Beziehung zu setzen. Erst recht gestattet sie, das Gewicht der Banalität und des Niederträchtigen im politischen Leben, das Ausgeliefertsein an Konstellationen deutlich zu machen.“*[2]

Dabei verbinden sich in einer Person die unterschiedlichsten Entwicklungslinien. Winfried Becker hat einmal betont, daß der Historiker

1 Vgl. *Jens Nordalm*, Der gegängelte Held. „Heroenkult“ im 19. Jahrhundert am Beispiel Thomas Carlyles und Heinrich von Treitschkes, in: HZ 276, 2003, 647–675.

2 *Heinz Gollwitzer*, Ein Staatsmann des Vormärz: Karl von Abel 1788–1859 (Schriftenreihe der Historischen Kommission bei der Bayerischen Akademie der Wissenschaften 50), Göttingen 1993, 12.

> *„die historische Persönlichkeit als einen Kreuzungspunkt mannigfacher geistiger, politischer und sozialer Handlungen und Geschehnisse“*[3]

betrachtet, in der sich eine Epoche widerspiegelt und oft in charakteristischen Zügen aufzeigen läßt.

Deshalb bin ich besonders glücklich, daß Karina Urbach und Brendan Simms die Rückkehr der großen Männer mit herausragenden Beispielen aus der britischen und deutschen Geschichte in den Mittelpunkt unserer Tagung gerückt haben. Ausgehend von der Rolle von zehn Staatsmännern im Krieg wurde in systematischer und länderübergreifender Perspektive die Rolle ziviler Führungskräfte im Krieg untersucht – und dies in einem Untersuchungszeitraum von 200 Jahren, von Friedrich dem Großen bis zu Winston Churchill.

Ohne die konzeptionelle Arbeit der Kollegen Urbach und Simms sowie die Bereitschaft der Referenten, die zentrale Fragestellung vor dem Hintergrund ihrer eigenen Forschungen zu beleuchten, wären die Konferenz und der vorliegende Tagungsband nicht möglich gewesen. So darf ich im Namen der Prinz-Albert-Gesellschaft den Organisatoren der Tagung, den Referenten und den Kollegen, die sich als Kommentatoren beteiligt haben, an erster Stelle besonders herzlich danken. Gleichzeitig gilt herzlicher Dank der Stadt Coburg, die gemeinsam mit der Niederfüllbacher Stiftung, der Sparkasse Coburg-Lichtenfels, der Universität und dem Universitätsverein Bayreuth das Zustandekommen dieser Konferenz ermöglicht hat. Vor der Konferenz fand auch im Jahr 2009 wieder ein begleitendes Studentenseminar „Skills for Historical Research in Britain“ statt. Seine Organisation und Durchführung haben die Herren Horst Gehringer und Oliver Walton übernommen. Für die Unterstützung bei der Durchführung der Tagung gilt besonderer Dank Frau Dagmar Hanke. Für die erneut bewährte Zusammenarbeit gilt dem De Gruyter-Verlag herzlicher Dank.

Die Prinz-Albert-Gesellschaft und die Herausgeber widmen diesen Band in Dankbarkeit für seine großen Verdienste um unser Anliegen ihrem langjährigen Vorsitzenden Prof. Dr. Franz Bosbach, der die Geschicke der Gesellschaft von 1995 bis 2008 leitete.

Bayreuth, im April 2010

Dieter J. Weiß

3 *Winfried Becker*, Georg von Hertling 1843–1919, Bd. 1: Jugend und Selbstfindung zwischen Romantik und Kulturkampf (Veröffentlichungen der Kommission für Zeitgeschichte B 31), Mainz 1981, 280.

Inhalt/Contents

Brendan Simms and Karina Urbach[1]

Introduction

Throughout late 2009 and early 2010, British politics was transfixed by the work of the 'Iraq Inquiry'. It was announced in June 2009 by Prime Minister Gordon Brown to examine "*the UK's involvement in Iraq, the military action and its aftermath*". The committee, made up of distinguished historians such as Sir Martin Gilbert and Sir Lawrence Freedman, and civil servants such as Sir Roderic Lyne and Sir John Chilcot, intended to

> "*learn lessons*" to "*ensure that if we face similar situations in future, the government of the day is best equipped to respond to those situations in the most effective manner in the best interest of the country*"[2].

Britain and the world remains bitterly divided about whether or not it was right to remove Saddam Hussein by force, but one thing is clear from the work of the inquiry: the role of the statesmen, particularly that of the then Prime Minister, Mr Tony Blair, in taking the country to war against strong domestic opposition was absolutely pivotal.

A historical consideration of the role of the statesman in war therefore seems timely[3]. With this in mind, the editors assembled a team of experts on British, German and international history to discuss the subject at Coburg in September 2009 under the auspices of the Prince Albert Society. The contributions ranged from an examination of the Seven Years war to the end of the Second World War. It would be wrong to impose a false coherence on our debates and conclusions. We know that the resulting volume cannot provide 'lessons learnt' in the sense of the Iraq Inquiry – always a risky proposition

---

1 The authors thank Tom Neuhaus and Matthew S. Seligmann for sharing their thoughts on the conference.

2 See Iraq Inquiry Website http://www.iraqinquiry.org.uk/about.aspx

3 The historical literature on this subject is, of course vast: e.g. *Eliot A. Cohen*, Supreme Command: soldiers, statesmen, and leadership in wartime, New York 2002, which appeared on the eve of the Iraq war and is said to have influenced President George W. Bush.

from the historian's perspective – because these lessons are retrospective and contextual, whereas the politicians seek something prophetic and universal. The editors share Herbert Butterfield's scepticism that history

> *"provides us with patterns which we can immediately transpose into the context of contemporary politics"*[4].

Like him, we believe that the study of history should promote wisdom, not mechanically but osmotically, and we echo his call for

> *"knowledge not to lie heavily on the mind, not to be used in a narrow or literal spirit, but to sink into the walls of [the historian's] brain so that it is turned into wisdom and experience- [allowing us] to acquire the right feeling for the texture of events, and would undoubtedly avoid becoming a mere slave of the past"*[5].

*

For a long time, the study of political leadership in peace and war was a central concern of British historians. One thinks, for example, of the mammoth early twentieth century biographies of Gladstone by John Morley[6], and Pitt, by Basil Williams[7]. Once the rash of studies celebrating Churchill and the national performance in the Second World War had passed, however, interest in leadership and biography waned considerably during the 1950s, 1960s and 1970s. This was partly to do with the rise of interpretative models – such as Marxism and Namierism – which were for obvious reasons largely unsympathetic to the idea of decisive personal agency in history. Less obviously, this hostility permeated even the 'high politics' school of history, often associated with the 'new right' and the Conservative Party. Historians such as Maurice Cowling, Alistair Cooke and John Vincent were too concerned to 'corrode' traditional pieties about great men and great events to pay much attention to the phenomenon of individual leadership[8]. Moreover, the historiography of the period 1945–1979 was very much focussed on wider issues such as imperial sunset and economic decline (managed or otherwise) with an implicit assumption that Britain was at the mercy of

4 *Herbert Butterfield*, History and human relations, London 1951, 173.
5 Ibid., 181.
6 *John Morley*, The life of William Ewart Gladstone, 3 volumes, London 1903.
7 *Basil Williams*, The Life of William Pitt. Earl of Chatham, London and New York 1913.
8 See *A. B. Cooke and John Vincent*, The governing passion. Cabinet government and party politics, 1885–1886, Brighton 1974; *Maurice Cowling*, 1867. Distraeli, Gladstone and revolution. The passing of the second reform bill, Cambridge 1967.

forces beyond her collective control let alone that of any particular leader. The biographies of her prime ministers – Eden, Macmillan and Wilson – were all in various ways studies of failure, or at least unrealised potential.

It may not be entirely coincidental that the arrival of strong premiers such as Margaret Thatcher and later Tony Blair re-ignited interest in the study of political leadership. Unusually for post-1945 Britain, both were front and centre not only in low intensity operations, but also in debating and executing conventional war. The appearance of Peter Clarke's *A question of leadership. Gladstone to Thatcher*[9] surely reflected the birth of new interest in the subject. Since then, there has been a steady stream of studies of political leaders, mainly but not exclusively biographical. Some of these have been written by non-academics, albeit on the basis of extensive primary research, such as Andrew Roberts's *Salisbury. Victorian Titan*[10] and Leo McKinstry's *Rosebery. Statesman in turmoil*[11]. At least five biographical studies of Churchill appeared in 2002 alone[12]. There has also been a revival in the political history of the monarchy[13]. An interesting phenomenon has been the rise of comparative biography, perhaps taking its cue from Alan Bullock's *Parallel Lives.* One thinks here of Richard Toye on Lloyd George and Churchill[14], and that by Richard Aldous on Gladstone and Disraeli[15].

Much of this interest, like the Iraq Inquiry, is driven by the desire to distil 'lessons' on leadership from the past, or at least to 'inform' current political debates and leaders. At the same time, the study of political leadership in Britain has always been shaped by the writings of the political class itself. We have already mentioned that John Morley wrote a biography of Gladstone. Winston Churchill penned one on his ancestor the Duke of Marlborough; and Roy Jenkins on both Gladstone and Churchill. More recently, there have been biographies of the anti-slave trading crusader William Wilberforce and William Pitt the Younger, by the former Conservative leader and current

9 *Peter Clarke*, A question of leadership. Gladstone, London 1991.
10 *Andrew Roberts*, Salisbury. Victorian Titan, London 1999.
11 *Leo McKinstry*, Rosebery. Statesman in Turmoil, London 2005.
12 *John Keegan*, Churchill, London 2002; *John Ramsden*, Man of the century. Winston Churchill and his legend since 1945, London 2002; Lord Moran's posthumous, Churchill at war 1940–45, London 2002 and *Klaus Larres*, Churchill's Cold War: the politics of personal diplomacy, New Haven 2002.
13 See *David Cannadine*, From reverence to rigour, *Times Literary Supplement*, 23.1.2004, 11–13.
14 *Richard Toye*, Lloyd George and Churchill. Rivals for greatness, London 2007.
15 *Richard Aldous*, The lion and the unicorn, London 2006.

shadow foreign secretary, William Hague,[16] and a biography of Robert Peel by the former Conservative Home and Foreign Secretary, Douglas Hurd[17]. These may have been in some ways derivative endeavours – Hurd in particular seems to have borrowed heavily from Norman Gash[18] – but they did bring to the subject a certain 'feel' for political practice which scholarly biographers may lack.

In America, Henry Kissinger, tried to explain to a wider audience the secrets of diplomacy by analysing leading policy makers. His monograph 'Diplomacy' covered 300 years and took a global look at foreign affairs – from Cardinal Richelieu to Mao Tse-tung[19]. There is no equivalent of a German politician writing on their predecessors. Though Chancellor Kohl was rumoured to be an admirer of Bismarck, he had to tread carefully, in case people drew comparisons between the 'two unifications' of Germany. There is also the general problem in Germany that biographies of statesmen are popular with the public but not highly regarded by the historical profession[20]. German historiography remains largely sceptical about the role of personality in history. After the somewhat uncritical celebration of figures such as Luther, Frederick the Great, and Bismarck before 1945, German historians since the Second World War, and especially since the 1960s tended to eschew biographical approaches. This was largely because the study of the individual was regarded as intellectually frivolous, and politically dubious. Structures, not individuals reigned supreme. There was widespread shaking of heads, for example, when John Röhl – a relentless critic of Kaiser Wilhelm II embarked in the late 1970s on a large-scale biography[21]. Attitudes softened a little when Lothar Gall wrote his celebrated biography of Bismarck[22], but the general disapproval remained; and when Sir Ian Kershaw brought out his magisterial two-volume study of Hitler[23], it was once

16 *William Hague*, William Pitt the Younger, London 2004, and William Wilberforce. The life of the great anti-slave trade campaigner, London 2007.
17 *Douglas Hurd*, Robert Peel. A biography, London 2007.
18 Mr Secretary Peel *(1961)* and The age of Peel *(1968)*
19 *Henry Kissinger*, Diplomacy, New York 1994.
20 For criticism of this: *Hans-Christof Kraus*, Geschichte als Lebensgeschichte. Gegenwart und Zukunft der politischen Biographie, in: *Geschichte der Politik. Alte und neue Wege*, Historische Zeitschrift, Beiheft 44, ed. by Hans-Christof Kraus and Thomas Nicklas, München 2007, 311–332.
21 *John Röhl*, Die Jugend des Kaisers 1869–1888, Munich 1993; *idem.*, Wilhelm II. Der Aufbau der Persönlichen Monarchie 1888–1900, Munich 2001; *idem.*, Wilhelm II. Der Weg aus dem Abgrund 1900–1941, Munich 2008.
22 *Lothar Gall*, Bismarck. Der weisse Revolutionär, Frankfurt 1980.
23 *Ian Kershaw*, Hitler 1889–1936. Hubris, London 1998, and *idem.*, Hitler, 1936–1945. Nemesis, London 2000.

against the 'structural' approach of the author, rather than the biographical detail which was applauded.

Recently German historians have started to debate whether there should be a 'Neue Politikgeschichte'.[24] At the Dresdner Historikertag in 2008 a special session was devoted to discussions of the 'new' versus the 'old' (i.e. diplomatic) concepts. The jury is still out on the question as to the form these new concepts should take, but it is encouraging that a new generation of German historians have realised how far behind their country is in comparison to its American and British colleagues when it comes to decisive sub-fields of political history like International Relations. That political history in a wider, global, setting could become interesting again has of course a lot to do with Germany's new role in the post 911-world, with the increasing economic, military and political demands and expectations that are made on it.[25]

The speakers were confronted with the following basic questions: What role have statesmen played in the outbreak of war? To what extent have they succeeded in dominating the high command? How easily have they stamped their personalities on the war and sustain it? To what extent have such statesmen become prisoners of the ruling circle? How have they coped with domestic opposition? And how important has the role of the individual statesman been in bringing such conflicts to an end?

The comparison between Germany and Britain proved very productive, nowhere more so than with regard to the epic Seven Years War (1756–1763). Tim Blanning's chapter on Frederick the Great shows that Prussian king's particular contribution to leadership in war was the concetration of all power and responsibility in his own person. He was the last true *roi connétable* in the traditional sense (unless one includes a usurper such as Napoleon Bonaparte). This made Prussia a formidable enemy, because Frederick could take risks, and take them without consultation, which no subordinate, collection of subordinates, or parliament could have matched. But if this made victory possible, it also greatly increased the chances of total defeat, which Frederick came close to on many occasions. In Britain, by contrast, Frederick's great ally William Pitt might also have been a risk taker,

---

24 *Ute Frevert*, Neue Politikgeschichte: Konzepte und Herausforderungen, in: *Neue Politikgeschichte. Perspektiven einer historischen Politikforschung*, ed. by Ute Frevert and Hein-Gerhard Haupt, Frankfurt am Main 2005, 7–26. Critical of her approach: *Hans-Christof Kraus, Thomas Nicklas*, Einleitung, in: *Geschichte der Politik. Alte und neue Wege*, Historische Zeitschrift, Beiheft 44, ed. by Hans-Christof Kraus und Thomas Nicklas, München 2007.

25 See for this: *Gregor Schöllgen*, Der Auftritt. Deutschlands Rückkehr auf die Weltbühne, Berlin 2004.

but Brendan Simms argues that his real contribution in his ability to forge and hold together a diverse domestic and popular coalition in support of a war effort which spanned several continents. One can argue whether this difference was attributable to the varying demands of strategy and geography or to fundamentally differing political cultures, or some combination of these factors, but of its existence there can be little doubt.

It would be quite wrong, however, to lay too much stress on the 'militaristic' nature of German leadership. In his brilliant examination of Joseph II, based on a lifetime's research for the definitive biography[26], Derek Beales shows the emperor to have been much less belligerent than often assumed. Instead of portraying him as the impractical dreamer of historiographical myth who charged all the domestic and diplomatic windmills at the same time, this chapter credits Joseph with a healthy sense of reality and *raison d'etat*. On many occasions, in fact, it was the emperor who had to restrain his hawkish chancellor Prince Kaunitz. There are therefore interesting parallels between him and the Younger Pitt, who is the subject of Frank-Lothar Kroll's contribution. He shows the British Prime Minister for much of the Revolutionary and Napoleonic period to have been a conservative, to be sure, but not a doctrinaire or inflexible one. By contrast with, say Edmund Burke, Pitt's objections were less fundamental and ethical than informed by a *realpolitik* sense that French power had exploded out of control and had to be contained.

This tension between ideology and the dictates of the international state system was also very much a feature of the two late nineteenth-century leaders examined here. The British Liberal Prime Minister William Ewart Gladstone has often been characterised as a protagonist of doctrinaire moral universalism in foreign policy and war. As Thomas Stamm-Kuhlmann shows in his chapter, there is a lot in this, particularly with regard to his involvement in the 'Bulgarian' agitation and his famous 'Midlothian' speeches. Yet, as the author also demonstrates, Gladstone's interventionist policy in Egypt – which he was wont to justify as an export of western values to the Middle East – sat uneasily with his proclaimed moral principles, especially as he stood to benefit financially from the deployment of British troops as a shareholder of the Suez Canal Company. Whether all this conspired to make Gladstone a potentially weak war leader, as some of the discussants believed, we shall never know for certain, because alone among all the statesmen ex-

26 *Derek Beales*, Joseph II. Volume I. In the shadow of Maria Theresa, 1741–1780, Cambridge 1987, and Joseph II. Volume II. Against the world, 1780–1790, Cambridge 2009.

amined in this volume, Gladstone never had to direct strategy in a major European war.

In this respect, the Liberal leader was very different from Otto von Bismarck, who navigated Prussia through a series of major conflagrations in the 1860s to unify Germany. He is traditionally portrayed as the epitome of German 'militarism' and Realpolitik. As Karina Urbach shows, this view is not completely wrong but it lacks crucial nuance. Part of the problem is that we have bought Bismarck's own 'myth', and that of later generations of nationalist celebrants. In his memoirs Bismarck portrayed himself as a rational leader, who had to keep the military under control. It is true that his wars were truly, to quote Clausewitz's famous dictum, a continuation of politics by other means, in that they were launched in carefully controlled diplomatic environments and although very bloody, were mercifully short. It remains an open question how Bismarck would have coped with a long struggle such as the Seven Years War, the Napoleonic Wars, the First World War or the Second World War. The indications are negative. When the Franco-Prussian War was in danger of becoming a long war, the Prussian leadership fell out. The fight between Moltke and Bismarck has been seen as the classic fight between the military and politicians. In fact both parties were interfering in each other's spheres – Bismarck portrayed himself as a soldier in the famous table talks and Moltke played with the idea to influence politics. That the situation did not get completely out of hand owes everything to the concept of monarchical legitimacy. The role of the third party in this war, William I, has often been underestimated. At the end of the day he was the decisive 'arbiter'.

It would be hard to find a more different figure from Bismarck – or Gladstone – than the great Liberal First World Leader David Lloyd George. Keith Robbins's chapter on him shows that despite his early pacifism, Lloyd George was the right man to lead Britain to victory over imperial Germany because of his capacity 'to get things done'. Whether this was in fact the case, has inevitably been questioned and qualified in the burgeoning secondary literature. In this context, Robbins's robust and persuasively-argued conviction that Lloyd George was 'the man of the hour' is a refreshing reminder of the importance of personality. The Kaiser, on the other hand, has gone down in history as a work-shy 'dilettante', to quote Max Weber's savage dismissal. Yet, as his latest biographer Christopher Clark has shown, Wilhelm II might not have been the omnipotent war leader of his own dreams, but he was by no means a cipher. Clark demonstrates how the emperor kept his fellow monarchs, the Kings and Arch-Dukes of the non-Prussian territories united behind the war effort. Moreover, it was the Kaiser's patronage which kept Erich von Falkenhayn as Chief of Staff in the face of furious sniping from the 'easterners' within the German military. One can argue whether this was the right decision, but there was nothing frivolous or lightweight about the resulting emphasis on the need to win the war through a decisive effort on the western front.

If the earlier contributions emphasized the importance of organisation, restraint, and various other factors, the two chapters on the Second World focussed on the issue of charisma. Klaus Larres showed how Churchill had the ability to captivate and motivate people. This was vital to mobilising British energies in the dark days of 1940 after the fall of France. As one of his cabinet secretaries later wrote

> *"A new sense of purpose and urgency was created as it came to be realized that a firm hand , guided by a strong will, was on the wheel"*[27].

By the end of the war, once the defeat of Hitler was certain, Churchill concentrated on parlaying his 'bulldog' spirit into concrete post-war advantage for Britain. For Hitler too, charisma was vital tool not only to mobilise German society for total war, but also to paper of the gaps in his own organisational and military skills[28]. In the discussion, Jonathan Haslam enlarged our understanding of Hitler's charisma to point out that no less a person than the Soviet leader Stalin, the most sceptical man in the world, had been taken in by Hitler's charisma, without ever even having met him.

Standing back, certain underlying themes emerge. Unsurprisingly, energy is right at the top of the qualifications needed to lead in wartime. The elder Pitt, who not only had to direct a war across several continents, but also keep together a fissile domestic coalition at home had it in spades. So did Joseph, who perhaps suffered from a surfeit of energy as he staggered from one domestic and external confrontation to another. The younger Pitt was tireless in welding together new confederations against Napoleon and mobilising vast financial resource. Similar qualities were exhibited by Bismarck, Lloyd George and of course Churchill. Energy, however, was useless, unless it was combined with judgment and a willingness to take calculated risks. Here Frederick, whose colossal gambles finally paid off, the Elder Pitt, who staked his reputation on the 'German War', Bismarck, who famously negotiated three military tightrope acts in 1864, 1866 and 1870, Lloyd George,

---

27 Quoted in *Gordon Craig*, Whose war is it?, *New York Review of Books*, 10.10.2002, p. 45

28 Unfortunately Professor Wolfram Pyta's paper on Hitler could not be submitted on time.

and Churchill, who refused to treat with Hitler even in 1940, come out well. But even Joseph II, who restrained Kaunitz from suicidal adventures and the Kaiser, who stopped Ludendorff from annexing Poland, were not completely without merit in this respect.

This raises, finally, the vexed question of political interference in military affairs, and the issue of civil-military relations more broadly. Frederick, of course, combined both functions in his own person so the problem does not arise. On the whole, both the elder and the younger Pitt were content to leave the execution of grand strategy to the generals and admirals. Bismarck, on the other hand, was constantly interfering, but did not always prevail. The same was true of Lloyd George. Perhaps surprisingly, the Kaiser chose to leave the day to day running of the war almost entirely to the High Command. Both Hitler and Churchill, of course, intervened manically. Sometimes they got it right, sometimes not. Churchill was probably justified in delaying the date of the D-Day landing, but the Greek campaign was certainly a major mistake. Likewise, Hitler could claim credit for sponsoring the famous Manstein Plan which felled France in 1940, but was responsible for many disasters including the Stalingrad campaign.

Whether or not a statesman had any personal military experience does not seem to have played much of role, either in their relations with the armed forces or with respect to the quality of their strategic decisions. The elder Pitt's brief service as a cornet of horse does not seem to have informed his direction of the Seven Years War. His son, the Younger Pitt, had no such experience at all, and yet managed to give Napoleon a run for his money until his untimely death in 1806. Bismarck, who had never seen action himself, fought three successful wars and never lost a campaign. Both Hitler and Churchill, on the other hand, had seen active service as young men, and yet it was they who could be most awry in their strategic judgments. The reason for this disjuncture is simple. It is in the nature of things that whatever knowledge of war the statesman has is likely to be picked up as young man in a relatively junior capacity (Frederick again is the exception here) and therefore of slender relevance to the direction of an entire war. That did not stop either Hitler or Churchill from lecturing their generals on the basis of their own experience. A little military knowledge can be a dangerous thing.

There are countless reasons for wars: religious, ideological, geopolitical, economic. Without overlooking these wider structures that provoke wars, the authors tried to draw attention back to the decision makers. In the end they judge whether these reasons are sufficient to go to war. The Iraq-inquiry confirms that even in a democratic state politicians can still make decisions – against the tide of opinion – that can change the course of history.

This book therefore hopes to bring back the role of personality into the wider debate. One of our core contributors, Derek Beales, adopted such an approach two decades ago:

> *"What I have tried to show is that biography has been too much disparaged. When a great historian can mistake a person for a trend, when it is thought more important to analyse social background than opinions, then the time has come for a reaction"*[29].

29 *Derek Beales*, History and Biography: An Inaugural Lecture 1980, in: *History and Biography. Essays in Honour of Derek Beales*, ed. by T. C. W. Blanning, Cambridge 1996, 266–283.

Tim Blanning

Frederick the Great

There could hardly be a clearer case of a statesman playing a decisive role in war than Frederick in 1740. He succeeded to the throne on 31 May and at once set about exploiting the formidable military and financial assets bequeathed to him by his father, Frederick William I. There is no reason to doubt Frederick's own candid admission that, first and foremost, he wanted to make a name for himself. As his troops marched into Silesia, he wrote to Voltaire that he would have behaved differently

> *„wenn mir das Phantom, das man Ruhm nennt, nicht allzu oft erschiene. Wahrhaftig, es ist ein großer Wahnsinn, aber einer, von dem man schwerlich loskommt, wenn man einmal davon ergriffen ist"*.

A little later, in February 1741, he wrote to his friend Jordan «*J'aime la guerre pour la gloire*», adding a few days later that he wanted to see his name in the newspapers and the history books[1].

On his accession he had canvassed the possibility of acquiring Jülich and Berg, but had got nowhere. It was then that the grim reaper came to his assistance, carrying off the Tsarina Anna of Russia on 23 October 1740 and the Holy Roman Emperor Charles VI three days later. On hearing the news, Frederick resolved "immediately" to "reclaim" Silesia. Writing of himself in the third person, he went on:

> *"this project accomplished all his political views; it afforded the means of acquiring reputation, of augmenting the power of the state, and of terminating what related to the litigious succession of the Duchy of Berg"*[2].

1 *Johannes Kunisch*, Friedrich der Große. Der König und seine Zeit, Munich 2004, 167.
2 *Frederick the Great*, The History of My Own Times, in Posthumous Works, vol. 1, London 1789, 92.

Silesia, he told Field Marshal Count Schwerin was

> *«l'objet le plus considérable qui s'est présenté depuis longtemps pour l'aggrandissement le plus solide, et le plus convenable à Son gloire et la grandeur de Sa maison»*[3].

His correspondence of late 1740 makes it clear that a major concern was that, if he did not take action, some other power might get in before him. He was especially concerned about Frederick Augustus, Elector of Saxony and King of Poland, who was married to a Habsburg and had a much better claim to a share of the Habsburg Monarchy than did Frederick. Silesia would provide the missing land bridge between Saxony and Poland and turn the Saxon-Polish complex into the dominant power in the region. In a memorandum entitled "Idées sur les projets politiques à former au sujet de la mort de l'empereur", written early in Nov 1740, Frederick wrote:

> *«Si nous voulions attendre pour agir que la Saxe et la Bavière fassent les premières hostilités, nous ne saurions empêcher la Saxe de s'agrandir, ce qui est cependant entièrement contraire à nos intérêts, et nous n'avons, en ce cas, aucun bon prétexte. Mais si nous agissons à présent, nous tenons la Saxe dans l'abaissement, et l'empêchant d'avoir les chevaux de remonte, nous la mettons hors d'état de pouvoir rien entreprendre»*[4].

The importance of the Saxon dimension in determining Frederick's actions has not always been given the attention it deserves. However obvious, the point is worth making that Silesia would be worth double its notional value to Prussia. If all the province's numerous assets – and there were plenty of them, for it was the most prosperous in the Habsburg Monarchy – were added together and expressed by the symbol '×', then its acquisition would change the power relationship by 2×, for what was acquired by Prussia was denied to Saxony or Austria. In footballing parlance, this phenomenon is known as a 'six-pointer'.

It might perhaps be argued that that a war over the Austrian succession was inevitable, given the parlous state in which Charles VI had left the Monarchy. Frederick himself told his representative in Vienna, Borcke, that it was

> *«sans chef, ses finance épuisées, son armée ruinée, ses provinces maltraitées par la guerre, la peste et la famine»*[5].

3 *Politische Correspondenz Friedrichs des Großen*, Vol. 1, Berlin 1879, 74.
4 Ibid., 90.
5 Ibid., 88.

Not only Saxony but Bavaria, whose Elector was also married to a Habsburg arch-duchess, had good claims and reasons to act. However, with a war in the West already underway between Spain and Great Britain, and with France almost certain to join in on the side of her Spanish ally, it was at least possible that the Austrian succession might have been settled by negotiation. It was Frederick's decision to strike that began a general war with truly world-historical consequences. It marked the beginning of Austro-Prussian dualism and, arguably, doomed the Holy Roman Empire. There could be no better illustration of the validity of Faust's translation of the first verse of the Gospel According to Saint John: "Im Anfang war die Tat!" Not only was Frederick his own prime minister and commander-in-chief, he was his own foreign minister too. No one could read through the documents of the *Politische Correspondenz* without appreciating that he enjoyed a degree of unity of command greater than any sovereign before or since. This was partly due to the relative simplicity of Prussia's civil and military structures, but mainly to his energy, application, intelligence and charisma. Such was the obedience he was able to inspire through a combination of fear and veneration, that he could raise his subordinates to an unprecedented pitch of exertion. They were also obliged to share his own nihilistic approach to death or glory:

> *„wenn alle meine Hilfsquellen, alle meine Verhandlungen versagen, wenn, mit einem Worte, alle Konjunkturen sich gegen mich erklären, dann will ich lieber mit Ehren untergehen, als verloren sein für mein ganzes Leben an Ruhm und Ruf«*[6].

Those words were written to Podewils on 27 April 1745, when the situation for Frederick was grim indeed. The campaign in Bohemia the previous autumn had ended in disaster, with his army straggling back to Silesia, decimated by disease and desertion. His French ally had lost interest in the war in Germany and was now concentrating on the struggle with the British. Frederick had also been abandoned by Bavaria, which had come to terms with Austria at the Peace of Füssen on 22 April. On 18 May Saxony signed an undertaking with Austria to assist in the recapture of Silesia. In Berlin plans were being made to evacuate government, archives and treasury to Magdeburg, Stettin and Küstrin, in case a defeat in Silesia enforced a last stand[7].

6 *Reinhold Koser*, Geschichte Friedrichs des Großen, 2 Vols., Suttgart und Berlin, 1904, Vol. I, 250.
7 Ibid., 249.

It was now that the advantages of unity of command became apparent. During the spring, he was "unusually busy", as Carlyle put it[8], organising new recruitment drives and ordering up reinforcements of men and matériel. Above all, by his constant presence in the main Silesian camps at Kamenz and Frankenstein he rebuilt the army and restored its morale[9]. By the middle of May 1745 he was ready for another round, so ready indeed that he deliberately left the Sudeten passes unguarded, to lure the Austrians and their Saxon allies into risking a decisive battle:

> *"Friedrich has left the mousetrap open; – and latterly has been baiting it with a pleasant spicing of toasted cheese"*[10].

His enemies took the bait and were defeated decisively at Hohenfriedberg on 4 June.

> *"So decisive a defeat has not been seen since Blenheim"*

he wrote to his mother[11].

The problem with unity of command, of course, is that it magnifies the chance of total victory – but also of total defeat. In this respect it is like charisma, as powerful as it is brittle. What was an asset at Hohenfriedberg (or Rossbach or Leuthen) was a serious handicap at when things went wrong. Only a field commander untroubled by responsibility to a higher authority could have persisted in pursuing so defective a battle-plan as he did at Kunersdorf[12]. His Austrian and Russian counterparts, inhibited by the need to obtain authorisation from Vienna or St. Petersburg, may have been slow-moving but at least they kept their armies intact. On more than one occasion, the *va banque* approach favoured by Frederick brought him to the very brink of disaster.

That he, his army and his state survived was due to his compensating ability to keep his aims limited. When Clausewitz delivered his central insight –

> *"war is not a mere act of policy but a true political instrument, a continuation of political activity by other means"*[13] –

---

8 'Friedrich in Silesia; unusually busy' was the title of chapter seven of volume four of his History of Friedrich II. of Prussia called Frederick the Great. 6 Vols., London 1858–60.

9 *Kunisch* (note 1), 210.

10 *Carlyle*, History of Friedrich II., Bd IV, 131.

11 Ibid., 132.

12 See, for example, the account of the battle in *Dennis E. Showalter*, The Wars of Frederick the Great, London 1996, 242–50.

13 *Carl von Clausewitz*, On War, ed. and trans. Michael Howard and Peter Paret, Princeton 1976, 87.

he had Frederick in mind. Frederick used military force to achieve precise objectives, primarily the acquisition and retention of Silesia, but he never allowed that force to become an end in itself. As Theodor Schieder wrote: 'if one understands by militarism the transfer of military ways of thinking to the political arena, then Frederick can be described as a militarist only with reservations, for his military actions were always determined by his political objectives'[14]. One consequence was that during his long reign Prussia was at war for less time than any of the other major powers.

It was this that distinguished him so sharply from the other great military commander of modern history to have enjoyed a comparable unity of command – Napoleon. In October 1814, Talleyrand said to the Archduke Johann:

> *"Napoleon never had a plan, it was always what had just happened that told him what to do next. If the Russian war had ended in victory, he would have gone on to Constantinople. He didn't know how to stop"*[15].

As Paul Schroeder has written with his usual perspicacity:

> *"all efforts to find some point in Napoleon's career when he turned wrong or went too far are misguided. His whole character and career were fundamentally wrong; he always went too far"*[16].

In his brilliant biography of Talleyrand, first published in 1932 and deservedly in print ever since, Duff Cooper made a particularly apposite comparison between Napoleon and Bismarck. More precisely, he compared their situation in the aftermath of total victory over Austria at Ulm and Königgrätz respectively:

> *"On each occasion Austria had been defeated, on each occasion the royal Commander-in-Chief was being urged on by his own ambition and by the voices of all his staff and of every soldier in the field to further triumphs, on each occasion the solitary, mistrusted voice of the Minister for Foreign Affairs, who alone could see in the defeated foe of today the potential ally of tomorrow, was raised in passionate pleading*

---

14 *Theodor Schieder*, Friedrich der Große. Ein Königtum der Widersprüche, Frankfurt am Main, Berlin and Vienna 1983, 218. Cf. *Gerhard Ritter*, Das deutsche Problem. Grundfragen deutschen Staatslebens, gestern und heute, Munich 1962, 27.

15 *Aus dem Tagebuche Erzherzogs Johanns von Österreich 1810–1815. Zur Geschichte der Befreiungskriege und des Wiener Kongresses*, ed. by Franz Ritter von Krones, Innsbruck 1891, 181.

16 *Paul W. Schroeder*, The transformation of European politics 1763–1848, Oxford 1994, 284.

for moderation. Bismarck succeeded where Talleyrand failed, but we know from Bismarck's own memoirs how near he came to failure and how the unexpected assistance of his old enemy, the Crown Prince, who came to his aid at a moment when he was contemplating suicide, turned the scales in his favour. Talleyrand did not take life or politics so seriously as Bismarck, but he saw the future as plainly, and once more he sought to dissuade Napoleon from the course that led to immediate triumph and ultimate destruction"[17].

Bismarck knew his history. In particular, he knew his history of Prussia in general and Frederick in particular. What he had learned from a close reading of Prussia's past was an awareness of the fragility of Prussia's geopolitical position, the constant threat from an encircling coalition and the need to avoid the sort of adventurism that had destroyed Napoleon I (and was to destroy Napoleon III too). As Walter Bussmann observed in the course of a lecture delivered at the Institut für Europäische Geschichte at Mainz in 1965:

"*especially in his own self-conception, Bismarck represented the Frederickian tradition*"[18].

If it is an exaggeration to claim that he was 'the last Prussian to be afraid', his Clausewitzian insistence on the need for limited aims, and the corresponding need to keep the military firmly under control, was increasingly at odds with the triumphalism of the younger generation.

That Frederick stamped his personality on the wars he fought cannot be doubted. He dominated representations of the conflicts in a manner that was both unprecedented and unparalleled – until the arrival of Napoleon Bonaparte. This was in large measure due to the unity of command already discussed. With the exception of George II, the last King of England to command his troops personally, at Dettingen in 1743, the other sovereigns were conspicuously absent from the battlefield. Although Joseph II took command of the Austrian army during the War of the Bavarian Succession of 1778, there was virtually no actual fighting in the course of this uneventful war of manœuvre dubbed derisively by contemporaries 'the Potato War' (because the troops spent most of their time foraging for potatoes).

It was not just that Frederick took command, it was the way he did it. From the first campaign of the War of the Austrian Succession, stories of his energy, decisiveness, self-sacrifice, courage – and a host of other related

17 *Duff Cooper*, Talleyrand, London 1932, 149.

18 *Walter Bussmann*, Otto von Bismarck. Geschichte – Staat – Politik, Wiesbaden 1966, 5.

military virtues – were in circulation. A special impact was made by his insistence that every last man in the army, from commander-in-chief to common soldier, should share the same privations and the same dangers. In no other army was the casualty rate among generals nor the rate of attrition among the military nobles so high. In the course of the three Silesian wars, 1,550 officers were killed in action, including sixty generals. The Kleist family lost twenty-three of its members, the Münchows fourteen, the Seydlitzes, and Franckenbergs and Schenkendorffs eight each, just to mention the most decimated[19]. Frederick himself narrowly escaped death on several occasions.

His heroic status was accentuated by the suddenness with which his wars catapulted Prussia into the top rank of European states-system. As early as 1745, at the end of the Second Silesian War, he was attracting the sobriquet 'the Great'[20]. The subsequent struggle to keep hold of what he had seized was elevated, for friend and foe alike, by the David-and-Goliath disproportion of the contest. Between 1756 and 1763 Prussia faced perhaps the most formidable coalition ever assembled in Europe: France, the Habsburg Monarchy, Russia, Spain, Sweden, and most of the German princes, against which he could only count on Great Britain, Hanover, Brunswick and Hessen-Kassel. In particular, the crushing victories over the French at Rossbach and the Austrians at Leuthen, in November and December 1757 respectively, furthered the image of superhuman potency.

Among the rich contemporary evidence attesting to his status, a special place should be found for the visual illustrations with which the period abounds. They ranged from sophisticated (and expensive) engravings such as Johann Georg Lorenz Rugendas's 'The triumphal return of Frederick to Berlin after the Peace of Hubertusburg 1763', or the numerous works of Daniel Chodowiecki, to simple wood-cuts sold by pedlars[21]. Images of Frederick and his generals also found their way on to consumer items such as enamelled snuff-boxes and silk sashes[22]. Paradoxically, Frederick's defeats – Kolin and Kunersdorf, for example – were also commemorated, as evidence of Our Hero's indomitable resolution in the face of adversity. In other words, Frederick stamped his personality on the conflicts at the level of popular perception.

---

19 *Schieder* (note 14), 65, 72.

20 Ibid., 480.

21 For excellent examples, see the exhibition catalogue Lovis Corinth "Fridericus Rex". Der Preußenkönig in Mythos und Geschichte, Berlin, 1986. The Rugendas engraving is on S. 58.

22 *Georg Schmidt*, Wandel durch Vernunft. Deutsche Geschichte im 18. Jahrhundert, Munich 2009, 168.

The wealth of eye-witness accounts makes it clear that he alone was in charge. Although he sought the advice of his veteran commanders such as Dessau, Ziethen, Winterfeldt and Schwerin, it was he who took the final decisions. In 1756 all of his closest advisers, including his two brothers Augustus William and Henry, were against a pre-emptive strike. Only Winterfeldt counselled action. It is possible that Frederick was swayed by him – certainly Prince Henry thought so – but it was only Frederick who could say to the hesitant Podewils:

«Adieu, Monsieur de la timide politique»[23].

At a more general level, however, it has been argued that Frederick was the prisoner of his nobles. The socially neutral absolutist state, standing above class interests, it is held, was a myth: everything Frederick did was in the long- or short-term interest of the Junkers. He himself wrote in his 'Political Testament' of 1752 that

"a sovereign should regard it as his duty to protect the nobility, who form the finest jewel in his crown and the lustre of his army"[24].

As good as his word, he put a stop to his father's campaign to expand his land-holdings at the expense of noble land-owners suspected of misappropriating portions of the royal domain. He favoured nobles for senior positions in the bureaucracy and throughout the officer corps. He restored to the Junkers their right to nominate candidates for the crucial position of Landrat. He ordered that no noble estates could be alienated from the class without his express permission. He established rural credit institutions to supply noble landowners with cheap and easy mortgages[25].

Moreover, as soon as Frederick tried to do anything of which the nobles disapproved, the limits of his power were exposed immediately. The Estates, the provincial organisations of the nobles, were still powerful and still able to ensure that the arm of the state reached only as far as they wished[26]. A

---

23 *Schieder* (note 14), 181.

24 *Die Politischen Testamente Friedrichs des Grossen*, ed. by G. B. Volz, Berlin 1920, 26.

25 *Wilhelm Bleek*, Von der Kameralausbildung zum Juristenprivileg. Studium, Prüfung und Ausbildung der höheren Beamten des allgemeinen Verwaltungsdienstes in Deutschland im 18. und 19. Jahrhundert, Berlin 1972, 69. *Karl Demeter*, Die Herkunft des preußischen Offizierkorps, in: *Moderne preußische Geschichte*, 3 Vols., ed. by Otto Büsch und Wolfgang Neugenbauer, Berlin und New York 1981, Vol. II, 875. *Peter Baumgart*, Zur Geschichte der kurmärkischen Stände im 17. und 18. Jahrhundert, in: *Ständische Vertretungen in Europa im 17. und 18. Jahrhundert*, ed. by Dietrich Gerhard, Göttingen 1969, 148. *Ingrid Mittenzwei*, Friedrich II von Preußen: Eine Biographie, East Berlin 1979, 155.

26 *Klaus Vetter*, Kurmärkischer Adel und preußische Reformen, Weimar 1979, 23–5.

particularly good illustration of the true location of power in Prussia was supplied by Frederick's attempt to abolish serfdom in Pomerania after the Seven Years War. After the Junkers there had rejected his plans emphatically, using the cogent argument that it was only serfdom which allowed them to serve in the army in such numbers, Frederick backed down[27]. At the centre, the reign saw a constant struggle between royal autocracy and the noble-dominated bureaucracy for supremacy. Frederick may have delayed the latter's victory – but he could not prevent it[28].

None of this is wrong, but neither does it present an adequate picture of Frederick's regime. Aristocratic by instinct and conviction, he also had a clear conception of the state that transcended class interests. When he wrote, as he often did, that he was 'only the first servant of the state', he did not mean that he 'only the first servant of the Junkers'. The reforms he promoted in the cultural, judicial, economic and religious fields demonstrated that his perspective embraced all members of his territories. Moreover, at the level of perception, his subjects did not feel that they were living in a state run by and for the Junkers. Moses Mendelssohn, for example, wrote:

> *"I live in a state in which one of the wisest sovereigns who ever ruled mankind has made the arts and the sciences blossom and a sensible freedom of thought so widespread that their effects have reached down to the humblest inhabitant of his dominions"*[29].

Frederick was able to ignore domestic opposition. So terrible was the loss of human life in the Seven Years War that there must have been much opposition from all classes. Frederick himself estimated that the war had cost Prussia half-a-million lives or around 10 % of the total population[30]. Yet there was no institutional means of expressing discontent. True, there was a rapidly expanding public sphere, but that was kept closely under control. At the beginning of his reign, Frederick had ordered that the Berlin press should be granted "*an unlimited freedom*", telling a sceptical official that

> "*if they are to be interesting, newspapers should not be obstructed*"[31].

It was of little significance that that order was simply not obeyed, for it was not long before Frederick himself changed his mind. By 1751 Lessing

27 *Klaus Epstein*, The Genesis of German Conservatism, Princeton 1966, 211–13.

28 *Hans Rosenberg*, Bureaucracy, Aristocracy and Autocracy: The Prussian experience 1660–1815, Cambridge, Mass. 1958, 155.

29 Quoted in *Ludwig Geiger*, Die deutsche Literatur und die Juden, Berlin 1910, 58–9.

30 *Horst Möller*, Fürstenstaat oder Bürgernation: Deutschland 1763–1815, Berlin 1989, 42.

31 *Ernst Consentius*, Friedrich der Grosse und die Zeitungs-Zensur, Preußische Jahrbücher, CXV (1904), 220.

could report from Berlin to his father that the newspapers there were sterile and tedious, on account of "*the strict censorship*"[32]. It was also Lessing who famously told the doyen of the Berlin Enlightenment, Friedrich Nicolai:

> "*don't talk to me about your Berlin freedom to think and write. It's just the freedom to market as many insults about religion as one likes*"[33].

Nevertheless, the fact remains that Frederick was untroubled by domestic opposition when it came to conducting the war.

Frederick would dearly have loved to have brought the Silesian Wars to an end just as soon as he had conquered the province. Unfortunately for him, he was confronted by a formidable foe in the shape of Maria Theresa, who was determined to regain Silesia at almost any cost. It was only after many years of immensely expensive and disruptive fighting that she could be brought to recognise that it was gone for ever. Even if she had been inclined to compromise earlier, there would still have been the interests and ambitions of all the other belligerents to be taken into account. As Clausewitz observed,

> "*everything in war is very simple, but the simplest thing is difficult*"[34].

That dictum applies with especial force to bringing wars to an end. In Frederick's case, perhaps a more apposite question would have been 'how important has the role of the individual statesman been in preventing conflicts in the first place?' For it was he who engineered the first partition of Poland in 1772 which almost certainly avoided a major war between the Eastern powers. As Hamish Scott has written, for him it was

> "*a considerable personal triumph*"[35].

Frederick's impact on war and diplomacy during his long reign was patently immense. Some of his more enthusiastic admirers believed that to call him 'the Great' was inadequate, for that was a sobriquet claimed for other rulers. Instead, they preferred "the Unique". In an important but value-free way this was correct. Even towards the end of his reign, the conduct of business on and off the battlefield was becoming so much more complex that a single individual could not master the flow of information. For a brief period, it looked as though Napoleon might manage it but it turned out that it was just what had made his conquests possible that condemned his regime to total failure. Frederick was lucky to have been born when he was.

---

32 Ibid., 230–1.

33 *Friedrich II, König von Preußen, und die deutsche Literatur des 18. Jahrhunderts*, ed. by Horst Steinmetz, Stuttgart 1985, 50.

34 *Clausewitz* (note 13), 199.

35 *H. M. Scott*, The birth of a great power system 1740–1815, London 2006, 166.

Brendan Simms

William Pitt the Elder, Strategic leadership at Home and Abroad during the Great War for the Empires (1756–1763)[1]

Until thirty years ago, the historiography of William Pitt's leadership during the Seven Years War was easy to relate. To the founding generation of strategic historians, Pitt was the presiding genius behind Britain's victory over the Bourbon powers in Europe and especially overseas. One thinks here particularly of the naval historians Alfred Thayer Mahan and Julian Corbett[2]. Over the past few decades our view of Pitt has become more nuanced. In the mid-1980s, Richard Middleton's study of conduct of the Seven Years' War put his role in the broader context of other major figures, especially his co-chief minister the Duke of Newcastle, but also Lord Anson at the Admiralty and Sir John Ligonier in the army[3]. In the late 1990s, Marie Peters challenged the notion of Pitt as the great imperialist, showing his interest in colonial expansion to have been belated and intermittent[4]. These interpretations in turn have been questioned by Karl Schweizer and Jeremy Black both of whom have in various ways insisted on the centrality of Pitt to British grand strategy in general, and on the strength of his interest in imperial expansion in particular[5]. This chapter is not primarily intended as a contribution to that debate. Instead, it tries to answer the central question

1 There is an extremely helpful, though now somewhat out of date, bibliography: *Karl W. Schweizer*, William Pitt, Earl of Chatham, 1708–1778, Westport/Conn 1993.

2 *A. T. Mahan*, The influence of sea power upon history, 1660–1783, Boston 1890, reprinted London 1965; *Julian S. Corbett*, England in the Seven Years War: a study in combined strategy, 2 Vols, London 1907.

3 *Richard Middleton*, The bells of victory. The Pitt-Newcastle ministry and the conduct of the Seven Years' War, 1757–1762, Cambridge 1985.

4 *Marie Peters*, The elder Pitt, London and New York 1998, and especially in The myth of William Pitt, Earl of Chatham, great imperialist: I. Pitt and imperial expansion, 1738–1763, Journal of Imperial and Commonwealth History 21 (2003), 31–74.

5 *Jeremy Black*, Pitt the Elder, Cambridge 1992. For a recent stirring account of the 'year of victories' see *Frank McLynn*, 1759. The year Britain became master of the world, London 2004. I have attempted to place Pitt in the continuity of British con-

about Pitt's leadership during the Seven Years' War: how it was that a man without any diplomatic or military experience, who lacked royal favour, managed to play such a decisive role in the defeat of France in Germany and America.

*

When Pitt became Secretary of State for the Southern Department in early December 1756, a post which he held with one short interruption, until October 1761, British fortunes were at a particularly low ebb. In North America, General Braddock's force had been massacred on the Monongahela in 1755. French regulars, and French-backed Indian tribes were pressing hard on British forts and settlements in the Ohio Valley, the Pennsylvania and Virginia backcountry and elsewhere in North America. A steady stream of French ships were slipping through the Royal Navy's blockade and heading west. In the Mediterranean, a French fleet had just captured Minorca and brushed off a half-hearted attempt by Admiral Byng to relieve it; a year later, the unfortunate commander was court-martialled and executed for this failure. As if all this was not bad enough, in May 1756 Britain's traditional Austrian ally had allied with France in the 'Diplomatic Revolution'. Hanover lay exposed; Britain was entirely friendless, with the exception of King Frederick of Prussia, thanks to the Convention of Westminster in January 1756, an ally or recent standing and as yet unproven value. To cap it all, Britain was rocked by an invasion scare at the end of the year, and had been forced to import German mercenaries to defend the south coast. The challenge confronting the new Secretary of State was immense. And yet by 1763, France had been contained in Europe, defeated in India and 'lost a continent' in America[6].

One of the greatest problems facing any leader, especially those in open or parliamentary systems, is their past record. Pitt did not come to power as a blank slate. Over the past twenty years he had spoken and acted in ways that could potentially cramp his style. Pitt made his foreign policy debut as a critic of Prime Minister Robert Walpole's Spanish policy in the 1730s[7]. He described the Convention of the Pardo (1739), a compromise arrangement

---

tinentalists in *Simms*, Pitt and Hanover, in: *The Hanoverian dimension in British history, 1714–1837*, ed. by Brendan Simms und Torsten Riotte, Cambridge 2007, 28–57; and in *Simms*, Three victories and a defeat. The rise and fall of the first British empire, 1714–1783, London 2007, esp. 422–462. This chapter draws on both works, but also incorporates some new manuscript material.

6 *Frank W. Brecher*, Losing a continent: France's North American Policy, 1753–1756, Westport/Conn. 1998.

7 *Black* (note 5), 7.

with Spain, as 'nothing but a stipulation for national ignominy'[8]. In the following decade, Pitt was at the forefront of those attacking the payments made to Hanoverian troops during the War of the Austrian Succession.

> "*It is now too apparent*", he told parliament, "*that this great, this powerful, this formidable kingdom, is considered only as a province to a despicable electorate*"[9].

Ten years later again, in the mid-1750s, Pitt criticised the subsidy treaties with Hesse, Russia and Prussia, as a

> "*chain and a connection, [which] would end in a general plan for the continent which this country would not support*"[10].

In short, Pitt took office with a reputation as an opponent of 'continental' measures in general and Hanoverian ones in particular. It is against this background that the development, execution and political 'sale' of Pitt's grand strategy needs to be viewed.

The overall framework of Pitt's strategy derived from classic Whig interventionism[11]. Over the next five years, he repeatedly stressed the need to safeguard the old Whig concerns with the 'liberties of Europe', the 'Protestant cause', with the European balance of power, in short. He spoke of the need

> "*to prevent the final overthrow of all Europe, and independency amongst the powers of the continent*", and of "*the formidable progress of the arms of France, and the danger to Great Britain and her allies resulting from a total subversion of the system of Europe*"[12].

The purpose behind British strategy, Pitt told the Commons was to keep Britain 'a Protestant, free, and an independent people'. This could only be effected, if Britons understood that that

> "*the interest of England … comprehends the whole system of public liberty in Europe*".

---

8 Quoted in Simms (note 5), 267.

9 Speech in William Cobbett, Parliamentary History, 10.12.1742, Volume XII, cols. 1033–6.

10 Pitt's remarks cited in Hardwicke to Newcastle, 9.8.1755, Powis House in *The life and correspondence of Philip Yorke, Earl of Hardwicke, Lord High Chancellor of Great Britain*, 3 vols., ed. by P. C. Yorke, Cambridge 1913, vol. II, 231.

11 For a view of Pitt's conception of the 'national interest' as more Tory, maritime and colonial see *The makers of British foreign policy from Pitt to Thatcher*, ed. by Jeremy Black, Basingstoke 2002.

12 Pitt to Keene, 23.8.1757, Whitehall, *Correspondence of William Pitt, Earl of Chatham* 4 Volumes, ed. by W. S. Taylor and J. H. Pringle (London, 1838–40), I, 248 and 251.

If that was "*shocked*" or "*endangered*", then "*the interest of England is endangered*". It was the same argument which had informed English foreign policy since the late sixteenth century. Moreover, he argued that if Britain remained "*an unconcerned spectator of the troubles of Europe*", France would have been in a better position to dispute 'the empire of the seas' with Britain, because Versailles would 'never be at a loss for materials to build a navy, while they preserve their superiority on the continent'. That being so, it followed that nothing could

> "*prevent France from having a marine superior to ours, whenever she is at freedom to bend her power and employ her subjects on that, and that alone*"[13].

The principal front, in short, was in Europe.

That meant Germany, for it was there that the French would have to be confronted on land.

> "*However inconvenient and expensive the German war is for England*", Pitt argued towards the end of the conflict in December 1761, "*it is more inconvenient and more expensive for France.*"

Pitt's determination to shape policy on central Europe was reflected in his demand during the coalition negotiations of late 1756 for the Northern Secretaryship, rather than the Southern department, covering France, Spain and colonial matters. Only George's stubborn opposition had denied him that office, and forced him to settle for the southern Secretaryship. All the same, Pitt now became the guiding spirit of Britain's commitment to Germany and remained so for most of the war. His very first speech as Secretary of State in the commons was in support of monies ostensibly towards the defence of Hanover. This paid for the Army of Observation, a Hanoverian force under the command of the Duke of Cumberland tasked with the defence of the electorate. Hanover was important in its own right, of course, but in Pitt's strategy it served chiefly as a bait to draw off French power on the continent, and a breakwater for Britain's principal ally Prussia. Pitt later exulted that

> "*France has beat herself in vain against the rock of Germany; she has found a resistance there that has exhausted her strength and she has been more shocked at the patience and magnamimity of His late Majesty' than by the victories of his fleets and armies in the Asiatic and American world*"[14].

13 Quoted in *Simms* (note 5), 425–26.

14 *Karl Schweizer*, A parliamentary speech by the Elder Pitt, Historical Research 4 (1991), 98–105.

Pitt was forced to show real leadership early on in the war, when the Commander of the Army of Observation in Germany, the Duke of Cumberland, concluded a neutrality agreement with the French at Kloster Zeven in September 1757. He immediately spoke of the "*fatal consequences*" that might

> "*follow a measure dictated by a precipitate spirit of selfish preservation*"[15].

Pitt was determined to tie down as many French troops there as possible and so he was at the forefront of those insisting as he put it

> "*that the English ministers must, for the sake of their honour, disculpate* [sic] *themselves*",

from the Convention[16]. There was a perceptible sigh of relief from Pitt when George finally repudiated the Convention and his son in the autumn. He wrote to his wife that

> "*The Army of Observation was before Hamburg, on the Elbe, and operations actually begun*"[17].

In short, support for the Army of Observation and for Frederick – in whom 'public liberty' of Europe now had 'a champion' – was central to the maintenance of British liberties and the maritime security on which they depended.

Expansion in America was part of this broader strategy. Its purpose was to secure the resources to sustain Britain's position in Europe, and to deny these to the French. This was, as the Commons was informed in 1757, a zero-sum game:

> "*The French … know that the source of power lies in riches, and that the source of the English riches lies in America. They know that in proportion we are weakened there, in the same proportion they are strengthened*".

The best way of countering this threat, the argument went, was

> "*by supporting his Prussian majesty, divert the French from pursuing those schemes that must infallibly give them a superiority by sea*";

15 Pitt to Lord Barrington (Holdernesse), 17.9.1757, Hayes, British Library (BL) Egerton 3432 (Leeds Papers), fol. 166.

16 Pitt's remarks are quoted in Newcastle to Hardwicke, 10.9.1757, Claremont, in: *Hardwicke* (note 10), III, 173.

17 'Minute of a meeting at Sir Conyers d'Arcy's lodgings', 7.10.1757 [at which Pitt, Newcastle and other important figures were present], National Archives (NA) 30/8/89, fol. 52. Pitt to Lady Hester Pitt, 1.12.1757, in: *Chatham Correspondence* (note 12), II, 2.

thereby *"we provide the most effectual manner for our settlements in America"*. In short, *"the distinction between Germany and America"*, was too narrow, *"for when they have nothing to fear from Germany, we shall have everything to fear for America"*[18].

This strategy rested on ensuring that no further enemy entered the lists on either side of the Atlantic. Here the principal worry was Spain. This was partly a naval issue, because the combined Franco-Spanish fleets would pose a threat to Britain's command of the seas. Pitt was also concerned, however, about the Spanish threat in Europe. The presumptive heir to the throne was the rabidly Anglo-phobe Charles of Naples. Pitt feared that Charles – who had refused to sign up to the Treaty of Aranjuez in 1752 – would try, as he put it, to secure

> *"to his second son the eventual succession to the kingdom his Sicilian Majesty now enjoys, in case he shall hereafter come to mount the throne of Spain"*,[19]

rather than handing it on to his brother Don Philip. This raised the spectre of an eventual re-union of the crowns of Spain and Naples and thus the resurrection of the Spanish empire in the Mediterranean. All this shows that even as Pitt was immersing himself in the colonial struggle and the German war, he remained sensible to the geopolitical complexities of the Mediterranean balance.

Initially, Pitt's strategy towards Spain was one of conciliation, even appeasement. Madrid had to be hugged close for as long as possible. In order to ensure this, Pitt was willing to think radically. He offered to surrender Gibraltar to Spain in return for Minorca, which had been occupied by France since mid-1756. To a later generation of navalists, this was tantamount to treason. An astonished Alfred Thayer Mahan – the renowned naval historian – was to comment that

> *"It seems incredible that even the stern and confident William Pitt should, as late as 1757, have offered to surrender to Spain the watch-tower from which England overlooks the road between the Mediterranean and the Atlantic, as the price of her help to recover Minorca"*[20].

Yet Mahan should not have been surprised. Minorca was, after all, regarded as the more important of the two bases throughout the eighteenth century.

18 Quoted in *Simms* (note 5), 426.

19 Pitt to Sir Benjamin Keene, 23.8.1757, Whitehall, in: *Chatham correspondence* (note 12), 253–54.

20 *Alfred Thayer Mahan*, The Influence of sea power upon history, 1660–1783, Boston 1890, reprinted London 1965, 298.

In short, for Pitt the question of Gibraltar was interpreted in the context of a wider British western Mediterranean strategy and global situation. If Gibraltar was the price for securing naval superiority overseas and at home, so be it.

Pitt also maintained a watching brief on the other neutral power: the United Provinces. Here there was a danger that the Royal Navy's highhandedness might drive them into the arms of France. At a time when Britain was in the midst of a new French invasion scare, the last thing she needed was another enemy, especially Holland, whose naval strength and geographical situation accentuated the threat to the British Isles. So in May 1759, Pitt was instrumental in pushing through an Act of parliament to regulate British privateers, much to the disgust of the Tories and British 'navalist' opinion[21]. All this suggests that if Richard Middleton was probably right to suggest Pitt's role in the global struggle has been somewhat exaggerated, he has perhaps not been given enough credit for his grasp of the implications of continental European politics for British strategy.

*

What is striking about Pitt's strategic leadership was his willingness to rethink and abandon entrenched positions not only those of his others but also his own. He had shown this even before taking up his war ministry at the end of 1756. In the summer of 1747, for example, Pitt rose above his earlier anti-Hanoverianism to praise the bravery of

> *"British and electoral troops [who] did all that can be expected from men overpowered by numbers"*

at the battle of Lauffeldt[22]. Three years later, he took a step back from his noisy Hispanophobia on the occasion of the Anglo-Spanish trade treaty of October 1750. As one observer reported,

> *"Pitt declared for a peace on any terms... recanted his having seconded the famous question for the 'no search'... said it was a mad and foolish notion, and that he was since grown ten years older and wiser".*

21 See *Alice Clare Carter*, The Dutch Republic in Europe in the Seven Years' War, o. O. 1971, 84–128, and *Richard Pares*, Colonial blockade and neutral rights, 1739–1763, Oxford 1938, 242–279.

22 Pitt to Richard Grenville, 30.6.1747, London, in: *The Grenville papers the correspondence of Richard of Richard Grenville Earl Temple and the Right Hon. George Grenville, their friends and contemporaries*, ed. by William James Smith, London 1852, 66.

In short, strategic imperatives had persuaded Pitt to set his commercial enthusiasm for confrontation with Spain to one side[23]. And in 1758, at the height of the Seven Years War, he freely admitted that

> "*he had formerly notions that this country could stand by itself and ought not to meddle with the continent*". Now however, "*he saw his error, it was a narrow and erroneous way of thinking*"[24].

A crucial quality among great leaders is the ability to show restraint where appropriate. After the hard-fought victory at Minden in August 1759, Pitt hoped that "*happy peace* [may] *wind up the glorious work, and heal a bleeding world*". Two months later, he told Hardwicke of his hope for a "*solid and general pacification of Europe*"[25]. In so doing, Pitt showed no hesitation in facing down the King's demands for an annexationist peace.

> "*Mr Pitt*", Newcastle reported in October 1759, "*had much ridiculed the King's way of talking about the conditions of peace and the retaining 'all our conquests'.*"

True, Pitt wanted to keep Niagara, Crown Point and various other territories as security for British colonies, but – the account continues –

> "*Quebec, Montreal and even Louisburg, they were points to be treated upon- not be given up for nothing, but what might deserve consideration and be a proper matter of negotiation*"[26].

Pitt, in other words, was prepared to make the hard choices necessary for peace.

On the other hand, Pitt did not shy away from drastic action when he thought it necessary. He watched the steady deterioration of relations with Spain after the accession of Charles III in 1759 with concern, and became convinced of a Bourbon conspiracy to attack Britain. In late August 1761, Pitt spoke of his fears that France and Spain were secretly planning a "*a naval union against England*" and "*forming a common nursery of seamen for that purpose*"[27]. By the autumn, he was calling for a pre-emptive strike on Spain before it could unite with France.

---

23 Cited in *Black*, (note 5), 84.
24 Quoted in *Simms* (note 5), 451.
25 Pitt to Lady Hester Pitt, 6.8.1759, in: *Chatham correspondence* (note 12), II, 9; Pitt to Hardwicke, 20.10.1759, Hayes, in: *Hardwicke* (note 10), III, 241.
26 Pitt's remarks are cited in: Newcastle to Hardwicke, 31.10.1759, Claremont, in: *Hardwicke* (note 10), III, 241–42.
27 Pitt to Stanley, 27.8.1761, Whitehall, BL Add Mss 57830, fol. 105.

> The *"total union of councils and interests between the two monarchies of the House of Bourbon"*, he wrote in a famous memorandum to King and cabinet, *"are matters of so high and urgent a nature as to call indispensably on His Majesty to take forthwith* [the] *necessary and timely measures"*[28].

He planned, as George Lyttleton remarked, to act à la Prusienne *and strike first, while the enemy was unguarded"*[29].

Here Pitt's strategy was not merely tactically pre-emptive, but also strategically preventive in conception. The main threat from Spain lay where it always had: in Europe. For contrary to the provisions of the Treaty of Aix la Chapelle, and as Pitt had feared, Charles did not leave Naples and Sicily to his brother Philip – thus ensuring the separation of those territories from Spain – but to his younger son Ferdinand. This maintained the status of Naples and Sicily as a Bourbon secundogeniture, but the separation was less final than if Don Philip had succeeded. It was perceived as keeping open the possibility of Ferdinand succeeding in Spain when Charles died, thus resurrecting the Spanish Mediterranean Empire, which had been broken up at Utrecht, and which Elizabeth Farnese, Philip V's second wife, had so desperately tried to revive after 1713. Pitt had been anxious about this for some time, and in mid-September 1759 he repeated his concern that

> *"the King of Naples might find himself at the head of the Spanish monarchy, and consequently possessed of all that due weight inseparable from that crown, which cannot but greatly affect the system of Europe"*[30].

In other words, Pitt was already looking well beyond the colonies towards the European balance of power itself, which would be upset if Spanish demands for Gibraltar and Minorca were realised, and perhaps even more so if the re-union of the Spanish Mediterranean empire came to pass through dynastic convergence[31]. Once again, we se Pitt's strategic leadership being informed by a deep grasp of European issues.

Pitt was also determined to prevent the re-emergence of French naval power further down the line. He told George III in mid-April 1761 that Britain could fight on until the "*total destruction of the French*" overseas[32].

---

28 'Advice in writing', 18.9.1761, in: *The Grenville papers* (note 22), I, 387.

29 Quoted in *Black* (note 5), 224.

30 Pitt is cited in *Zenab Esmat Rashed*, The peace of Paris 1763, Liverpool 1951, 46.

31 'Advice in writing', 18.9.1761, in: *The Grenville papers* (note 22), 387.

32 As reported in Newcastle to Hardwicke, 17.4.1761, Newcastle House, in: *Hardwicke* (note 10), III, 315.

He saw a unique opportunity to break the back of the French navy once and for all. Ejecting the France from Canada settled control of North America but it left the global task half-done. Pitt in short, was seeking to pre-empt the emergence, or re-emergence of a strategic threat many years down the line. In 1760 he demanded the cession of

> *"all Canada, Cape Breton, the islands, the harbours, the fisheries, and particularly the exclusive fishery of Newfoundland".*

After all, he argued, France was 'chiefly, if not solely, to be dreaded ... in the light of a maritime power'. For this reason it was necessary *"to break beyond hope of resuscitation the naval power of France"*. When peace was finally agreed at Paris in 1763, Pitt objected to the failure to retain the exclusive rights to the Newfoundland fisheries. As it turned out, Pitt was right to fear a French *revanche*. Already in early April 1762, even before peace had been concluded, Choiseul was writing that

> *"If I were in charge ... England would be reduced and destroyed by us within thirty years ... our navy is in bad shape, but it will recover, not everything is done in a day".*

The course of the American war more than a decade later, in which a resurgent French navy played such an important role, vindicated Pitt's stance in many respects[33].

*

The quality of Pitt's leadership was evident not merely in the conception, but also the execution of policy. No doubt his directing mind has been somewhat exaggerated by subsequent mythmakers, but informed contemporaries were in no doubt about the dynamism and cohesion which Pitt brought to the British war effort. The Austrian chancellor Kaunitz, at this point allied to Britain's enemy France, described in May 1758 how

> *"Pitt has his hand on the tiller. His former conduct has furnished convincing proof that he does not live from day to day, but goes to work in a systematic fashion, and know how to direct the decisions in his part of the world towards a single aim"*[34].

The Duke of Newcastle, no uncritical admirer, was to remark later on in the war that

33 Cited in *Rashed* (note 30), 134.
34 Quoted in *Black* (note 5).

> *"I know nobody who can plan or push the execution of any plan agreed on in the manner Mr Pitt did"*[35].

No wonder he recommended Pitt to a reluctant George II as the "*only one who has the ability to do this business*", including the "*German business*" which was so close to the King's heart[36].

It is worth restating just what an achievement Pitt's direction of the war effort was. To be sure the tactical implementation of the grand strategy was left to the commanders and the quartermasters, but each operation had to be conceived and planned in negotiation with a bewildering array of departments and political egos, not least that of his principal collaborator in government the Duke of Newcastle. The activities of the two branches of British foreign policy, the Northern and the Southern Departments had to be coordinated as far as possible. Ships and men had to be allocated and marshalled across the globe, from Canada, to the Caribbean, to Germany, the coasts of France and many other places. Soldiers and sailors had to be persuaded to make war on the enemy rather than each other: not an easy task given the propensity of what later generations have termed 'combined operations' to end in recrimination between army and navy[37]. Prickly commanders had to be massaged, and if necessary dealt with; of this more later. Militia commissions had to be distributed and courts-martials organised[38]. He had to cope with the King, and with Britain's no less tempestuous allies; of this too, more later. All the while, Pitt, as his correspondence shows, was frequently afflicted by gout and other ailments.

Equally crucial was Pitt's ability to cooperate with and manage a series of allies and auxiliaries. The key to his success in North America was the fresh dynamism he injected into the war there, and the new relationships he established with the colonists. In the past, colonial warfare had been conducted very much on *ad hoc* basis, with various coalitions of willing American militias supporting British regulars and the Royal Navy. Pitt now greatly increased expenditure on the American war, stepped up the number of British line infantry to be sent across the Atlantic, and eventually per-

---

35 Newcastle to Hardwicke, 15.10.1761, Claremont, in: *Hardwicke* (note 10), III, 338–339.

36 Newcastle to Hardwicke, 14.10.1756, in: *Hardwicke* (note 10), II, 323.

37 On this see *Richard Harding*, Soldiers and gentlemen of parade. Some professional and technical problems concerning the conduct of combined operations in the eighteenth century, Historical Journal 32 (1987), 35–55.

38 Pitt to Holdernesse, 18.5.1759, Whitehall, BL Egerton 3432, fol. 203, re the appointment of officers for the North Riding of Yorkshire; Holdernesse to Barrington, 3.12.1757, Sion Hill, BL Egerton 3432, fol. 29 describes how the 'whole of the transaction [concerning the Mordaunt court martial] passed through his hands'.

suaded parliament to pay for local levies. As Fred Anderson has pointed out in his magisterial *Crucible of war*, Pitt made the colonists full partners in his imperial enterprise[39]. He was also the driving force behind the appointment of a British supreme commander for North America, the Earl of Loudoun. Finally, Pitt was supportive of diplomatic efforts to win over the Indian tribes, most of whom had previously tended to side with the French[40].

Pitt's principal allies, however, were to be found in Europe. Here two figures were crucial to the war effort: Ferdinand Duke of Brunswick, the extremely capable commander in chief of His Britannic Majesty's Army in Germany, and Frederick the Great of Prussia. At Pitt's urging – though not only his – London contracted to maintain a large force of men in Germany, promising as many as 50,000 by 1758. Two months later, the subsidy to Hesse-Cassel was renewed. At first this army was made up of Hanoverians and other German levies, but in June 1758, Pitt took advantage of the parliamentary recess to propose the dispatch of 6,000 infantry and four cavalry brigades to Germany. This was a major new departure, rendered even more controversial by the inclusion of infantry, which could otherwise be deployed in North America or on coastal operations. From then on nearly a third of Ferdinand's army was to be made up of British troops[41].

Managing Ferdinand was a diplomatic and domestic minefield for Pitt. If the Duke himself was touchy, the senior British officers serving under him were no less so. In August 1758, for example, the Duke of Marlborough, who commanded the British contingent in the Duke's army, was complaining that a Hanoverian "*had been made a general of foot over my head*". This prompted the Duke to demand an equivalent.

> "*I hope*", he told Pitt with asperity, "*if I am thought unworthy of that rank personally, I may be excused, as an Englishman, for not quite tamely submitting to so strong a mark of the English being thought fit for nothing, but to be hewers of wood and drawers of water to the Hanoverians*"[42].

---

39 *Fred Anderson*, Crucible of war. The Seven Years War and the fate of empire in British North America, 1754–1766, New York 2000.

40 On Indian diplomacy see *Matthew Ward*, Breaking the Backcountry. The Seven Years war in Virginia and Pennsylvania, 1754–1765, Pittsburgh 2003, 163 and 238 and *Wilbur F. Jacobs*, Diplomacy and Indian gifts: Anglo-French rivalry along the Ohio and Northwest frontiers, 1748–1783, Stanford 1950, 161–167.

41 The terms of the agreement are described in 'Declaration'. 11.4.1758, London, NA 30/8/89, fols. 74–75. See also *Wolfram Pyta*, Von der Entente Cordiale zur Aufkündigung der Bündnispartnerschaft. Die preußisch-britischen Allianzbeziehungen im Siebenjährigen Krieg, 1758–1762, FBPG, New Series 10 (2000), 1–48.

42 Duke of Marlborough, 18.8.1758, in: *Chatham correspondence* (note 12), I, 338.

That particular storm blew over, but another was whipped up in the aftermath of the battle of Minden, in August 1759, when Ferdinand effectively accused the commander of the British cavalry, Lord George Sackville, of cowardice, did not. In the past, Pitt had been a shrill partisan of British generals supposedly slighted by the Hanoverians, and his relations with Sackville hitherto had been very cordial. Now he unhesitatingly took the Duke's side. Sackville's biographer Piers Mackesy describes Pitt's correspondence with the British commander as being

> *"phrased in that unique fusion of manly honesty with Uriah Heep which makes one's skin crawl as one reads his letters"*[43].

Sackville was court-martialled, and some thought him lucky to escape with his life. Pitt's attitude had been decisive and it was driven by his over-riding imperative to stay on good relations with the Duke of Brunswick, Britain's greatest military asset in North Germany.

The other key relationship in Germany was with the Prussian King, Frederick the Great. Pitt cultivated this through the British envoy Andrew Mitchell, a close associate[44]. He was also an active proponent of the Frederick cult[45]. From the start of the war to his resignation in late 1761, Pitt supported Frederick both financially and psychologically. Thus in late March 1757, Pitt expressed his

> *"sentiments of veneration and zeal for a prince [Frederick], who stands the unshaken bulwark of Europe, against the most powerful and malignant confederacy, that ever yet has threatened the independence of mankind"*[46].

At the same time, Pitt became a vocal champion of the Prussian connection – it was never a formal alliance, although George referred to Frederick as "*his good ally*" in his message to the House of Commons. This support did not go unrecognised in Berlin and stiffened Frederick in his resolve. A subsidy treaty with Prussia in April 1758 promised £670,000 in subsidies over four years; these defrayed a substantial proportion of Frederick's war costs, per-

---

43 *Piers Mackesy*, The coward of Minden. The affair of Lord George Sackville, London 1979, 163 (see also 42, 155, 162–164, 169).

44 See *Patrick Francis Doran*, Andrew Mitchell and Anglo-Prussian diplomatic relations during the Seven Years War, New York and London 1986.

45 On this topic more generally see *Manfred Schlenke*, England und das friedrizianische Preußen, 1740–1763. Ein Beitrag zum Verhältnis von Politik und öffentlicher Meinung im England des 18. Jahrhunderts, Freiburg and Munich 1963.

46 Pitt to Mitchell, 31.3.1757, Whitehall, in: *Chatham Correspondence* (note 12), I, 226.

haps as much as a quarter[47]. In 1760, Pitt was still publicly comparing him to Alexander the Great, "*who had only one empire to fight*"[48].

Pitt was prepared to take considerable risks to keep Frederick in the war. He promised to undertake "*useful diversions*" on the French coast, launching expeditions in 1757–1758 and again in 1760. Even more remarkably, Pitt refused to sell his ally short through a separate peace, and did not hide his willingness to forego colonial gains in order to shield Frederick. In December 1758, he refused to guarantee Tory demands for a guarantee that the vital Canadian fortress of Louisburg would never be returned to the French:

> "*he would not give up an iota of our allies for any British consideration*"[49].

Pitt also argued that the maintenance of the Prussian alliance was worth the sacrifice of colonial gains. When Newcastle told him in April 1760 that fiscal constraints meant that the war would have to be brought to a close within the year, Pitt responded

> "*that we did not want a peace, but that for the sake of the King of Prussia, we were willing to forego the great advantages which we had reason to promise ourselves from this campaign in every part of the world*"[50].

Indeed, Pitt insisted that the Prussians were kept fully abreast of British intentions in meetings in September 1759 and again in March 1760. In July 1761, he re-iterated his insistence on

> "*so capital and indispensable a condition of compensation, as the immediate restitution and evacuation of the places and territories conquered by France from his majesty's allies in Germany and… those of the King of Prussia*"[51].

Pitt was also highly skilled at building a domestic base for his war strategy. First and foremost, he had to win over or at least mollify the King, who had never forgiven him for his outspoken attacks on Hanover in 1740s. At first, relations with the King remained poor. Throughout the first half of

---

47 For Frederick's gratitude see the letter of the British envoy to Prussia, Andrew Mitchell: Andrew Mitchell to Pitt, 12.3.1757, Dresden, in: *Chatham Correspondence* (note 12), I, 225.

48 Pitt to Fredrick, 'November 1760', in: *Chatham Correspondence* (note 12), 84.

49 Quoted in *Marie Peters*, Pitt and popularity, Oxford 1980, 133.

50 As reported in Newcastle to Hardwicke, 9.4.1760, Newcastle House, in: *Hardwicke* (note 10), III, 244.

51 Pitt to Bristol, 28.7.1761, Whitehall, BL Add Mss 57830, fol. 169.

1757, the monarch continued to regard Pitt as an intolerable imposition and dismissed his commitment to the defence of Hanover. In April 1757, Pitt was unceremoniously sacked, at the behest of Cumberland, who wished to cover his back before setting off for Germany. There was then a popular clamour for his return, which unsettled the crown and the establishment. All attempts to construct an alternative ministry failed and George was forced to back down and re-instate Pitt. In time, however, the King came to appreciate Pitt's commitment to the German war and by the autumn of 1757 they were seeing eye to eye on the need to repudiate the Convention of Kloster Zeven. Thereafter, an uneasy truce between them was established which lasted until the monarch death in 1760.

Pitt's popularity was never as great as he claimed[52]; all the same, his ability to appeal to a broad national constituency was crucial to his success as a war leader. He depended heavily on parliamentary and press allies to 'sell' his strategy to the public. Here his principal supporters were the flamboyant London MP William Beckford, and his brother Richard, publisher of the widely-read Tory weekly essay paper *Monitor*[53]. By early November 1756, Beckford had become so alarmed at the "*melancholy prospect of public affairs*" – by which he meant the loss of Minorca and the retreat overseas – that he turned directly to Pitt as

> "*the instrument of our deliverance'. A new system is now absolutely necessary*", he added, and "*in our present political warfare, I intend to act as one of your private soldiers without commission*"[54].

As the former Secretary of State the Earl of Chesterfield observed in January 1758,

> "*Everything goes smoothly in Parliament; the King of Prussia has united all our parties in his support; and the Tories have declared that they will give Mr Pitt unlimited credit for this session*"[55].

In 1759, the Under-Secretary Richard Potenger remarked that thanks to the recent victories

---

52 For a sceptical view of Pitt's popularity see *Paul Langford*, William Pitt and public opinion, 1757, English Historical Review, 88 (1973), 54–79.

53 See *Marie Peters*, The *Monitor* on the constitution, 1755–1765: new light on the ideological origins of English radicalism, English Historical Review, 86 (1971), 706–727.

54 William Beckford to Pitt, 6.11.1756, Fonthill, in: *Chatham correspondence* (note 12), I, 185–6.

55 Chesterfield is cited in *Clark*, Dynamics of power, 455.

"every thing passes in the quietest manner in the House of Commons; there is scarce more than one voice there upon affairs of the most national consequence"[56].

Inevitably, the vigorous prosecution of the war in Europe dented Pitt's popularity. He had inherited a poisoned chalice. Controversial decisions about the war, particularly Europe and Hanover, would have to be made. Yet Pitt could not risk his greatest asset: his popularity and reputation for 'patriotic' probity. As Robert Walpole's son Horace remarked with some glee in November 1756,

"If he Hanoverizes, or checks any inquiries, he loses popularity, and falls that way: if he humours the present rage of the people, he provokes two powerful factions"[57].

For some, even the acceptance of office was a betrayal. *The Test* lampooned

"that ever-memorable Patriot, who after the mountain had groaned for a considerable time, was at length delivered of his mouse ... His mighty promises and all his patriot cares, which in the morning dazzled the reader's eye, perished, like summer insects, with the setting sun, and have ever since been chilled and inanimate".

It too reminded readers of Pitt's "*unembarrassed countenance*" in the 1740s, when he had violently opposed the Hanoverian troops "*in one session of parliament*" only to reverse himself "*the very next year*"[58].

As the war dragged on, and the commitment to Germany escalated, many in the public sphere turned against Pitt. A widely-discussed article in the *Gazetteer* in the summer of 1758 did not object to subsidizing Prussia but claimed that most would

"repine at the sending of forces to Germany, because they imagine it is done for the sake of Hanover, and fancy they perceive something unconstitutional in it".

Even the friendly *Monitor*, shaken by the dispatch of British troops, fired a broadside against their hero in July 1758[59].

"*I convinced the House of Commons that the troops sent to Emden were not designed for Germany*", one anti-Pittite satire ran, "*I made anti-Hanoverians and the enemies of Germany send men we could ill*

56 Quoted in *Jeremy Black*, Parliament and foreign policy in the eighteenth century, Cambridge and New York 2004, 245.

57 Quoted in *Peters* (note 49), 66.

58 *The Test*, 1, 6.11.1756, p. 1, 4.

59 Cited in *Peters* (note 49), 118–19.

> *spare, and money we could ill raise to the aid of Prince Ferdinand ... The oftener I altered my opinion the more I was esteemed for my consistency*"[60].

In 1761 one pamphleteer rejected Pitt's diversionary strategy on the grounds that

> "*the supplying a French army in Germany did not take one hand from one of the ports of France, their navy was not impaired by it, their marines and seamen were not lessened*"[61].

The maintenance of Pitt's leadership at home entailed a number of strategies to contain these attacks. One was to argue that his policies were, despite appearances, in fact completely in line with past strategic pronouncements, having being merely adapted to suit circumstances. As one observer reported:

> "*Mr Pitt moved for two hundred thousand pounds to be granted to the King for the support of an army of Observation in Germany to protect his Majesty's Electoral Dominions against any invasion from France. Mr Pitt upon this made a long speech said among many things that if any gentleman had a mind to attack his consistency that he hoped ... he did not insist on being literally & nominally consistent as long as he thought he was substantially so*"[62].

Another strategy was to distance himself from unpopular policies as far as possible.

Despite the fact that the dispatch of troops to the continent had been his initiative, Pitt put on a great show of reluctance. As a speechless Newcastle wrote in mid-July 1758,

> "*If you hear that Mr Pitt was against sending these troops abroad, it is entirely false. Nobody was more forward than himself; and three regiments were added ... without my knowledge tho' greatly with my approbation*".

There was ostentatious sucking of teeth and head-shaking at the costs, but these, as with the number of troops continued to rise[63].

---

60 Quoted in *Black* (note 5), 223.
61 Quoted in *Simms* (note 5), 471.
62 Digby to Ilchester, 1.3.1757, BL Add Mss 51341, fol. 42–43.
63 Newcastle to Henry Campion, 15.7.1758, Add Mss 32881, fol. 335. On this subject more generally see *P. F. Doran*, passim and *Karl W. Schweizer*, England, Prussia and the Seven Years war. Studies in alliance policies and diplomacy, Lewiston/Queenston/Lampeter 1989, esp. 62–63.

> *"It is amazing to me"*, the Lord Chancellor, the Earl of Hardwicke remarked in 1760, *"that considering how much depends upon the events of this campaign in Germany, he [Pitt] should not even propose the sending of some British infantry to that country. Perhaps he wants to be ravished to it"*.

This was a very astute observation: much of Pitt's public persona and personal popularity depended on the impression of being forced into controversial but necessary measures in Europe[64].

Another way of bolstering his standing with the public was by undertaking popular military operations. From 1757–1761, Pitt was to launch a series of coastal raids on the French Atlantic seaboard, including Rochefort, St. Malo, and Isle de France. They were intended to divert French forces from Germany, thus taking the pressure off Frederick's western flank. They would, as the *Monitor* noted in August 1757,

> *"presently cure [France] of marching the strength of their country ... beyond the Rhine"*[65].

The method might have been naval, but it was employed in pursuit of a double continental policy: France was to be distracted not only in Germany but on her coasts. These amphibious operations were to meet with decidedly mixed success, but they allowed Pitt to have it every way. They went some distance towards meeting Frederick's demands, they reduced pressure from local enthusiasts for the German war to send troops to Germany, they employed forces that were not currently deployed anywhere else. Best of all, the attacks could not be dismissed as a Hanoverian measure, they made the German subsidies more palatable, and appealed to all those who saw Britain's calling as a fundamentally maritime one.

By 1759, this inter-locking domestic and military strategy had taken on a dynamism of its own. Pitt used his popularity to wage war, while successful war increased his popularity. The breakthrough here was the capture of Louisburg by British forces in the summer of 1758. This success, which reversed the controversial cession of ten year before, gave Pitt's public standing an immense boost. No fewer than fifty addresses of congratulations followed, in addition to a wave of triumphant poems and ballads. Pitt immediately claimed this as a vindication of his strategy. It was a virtuous circle.

64 See Hardwicke to Newcastle, 10.4.1760, Moor Park, in: *Hardwicke* (note 10), III, 24–5.

65 Cited in *Peters* (note 49), 92. The connection between the coastal raids and the need to appease Frederick emerges very clearly from 'Substance of what passed between Mr Secretary Pitt & My Lord Holdernesse', 27.3.1758, BL Egerton 3432, fol. 176.

This brings us, finally to the most remarkable aspect of Pitt's leadership: the ability to take his followers with him to a new place. The parliamentary Tories and navalists who backed the subsidies for Frederick and the dispatch of British troops to Germany had only been partly bamboozled; in substantial measure they had also been persuaded. Pitt had created a new synergy between continental and colonial strategy, which rested on and re-inforced a domestic consensus. The Prussian subsidies encountered no Tory opposition in parliament in December 1758, and in January 1759, Beckford averred that the Prussian subsidy

> "*was giving money to America by this diversion on the continent of Europe and that if Hanover were lost, he would vote one hundred millions sooner than not recover it*"[66].

Truly, as Pitt was later to boast, Pitt had brought the Tories to support his war coalition on "*Whig principles*"[67], by which he meant, of course, foreign policy principles.

In broad sections of the public sphere, too, the conversion was complete. Here again, the *Monito*r led the way. Support for Hanover and Prussia, it argued repeatedly throughout the year, was not ministerial toadying, but part of a broader British project for the "*preservation of a bulwark to arbitrary power and universal monarchy on the continent*". 1759 also saw the creation of the London 'Troop Society' which was dedicated to help the widows and orphans of men killed in Germany. Loyal addresses to congratulate the King were already making fewer attempts to highlight colonial over continental victories. The two theatres were increasingly seen, even by former navalists, as part of a single war effort. Even more to the point, popular opinion had accepted the central role of Hanover. Thus one print showed the British lion in collaboration with the Hanoverian horse against the French cock.

> "*O pretty! O pretty!*", the lion exclaims, "*Thus hast save*[d] *me a great deal of labour and trouble, I have crush'd the Cock & secured America*"[68].

In effect, Pitt had created a new basis of legitimacy for himself. His standing in the country and particularly the City of London now rested on his conduct of the war rather than the patriot postures of previous years.

---

66 Quoted in *Peters* (note 49), 132.

67 The remarks about 'Whig Principles' are cited in *J. C. D. Clark*, The dynamics of change: the crisis of the 1750s and English party systems, Cambridge 1982, 455.

68 See *M. John Cardwell*, Arts and arms: literature, politic and patriotism during the Seven Years War, Manchester 2004, 246–47.

To sum up. Pitt demonstrated a high level of strategic leadership across the board during the Seven Years War. Conceptually, he coopted and refined the Whig strategic tradition of European interventionism which went back to the Glorious Revolution. Onto this he grafted elements of the antithetical 'blue water' strategy, with its emphasis on naval and colonial activism. In so doing, Pitt created – temporarily at least – a perfect strategic virtuous circle, in which Germany was defended in America, and America was won in Germany. He turned Hanover from a liability into a geopolitical asset; he created in Frederick the Great a Protestant hero to replace the 'old system' of alliance with Austria; and he managed multiple fronts and alliances with aplomb. He showed himself capable of both moderation and firmness, pleading for territorial restraint in some cases, and pushing for robust pre-emptive action in others.

None of these strengths would have amounted to much, had not Pitt been able to rally a motley army of radicals, planters, Tories and orthodox Whigs at home. Their support was vital to the pursuit of the war effort, which would have ground to halt without parliamentary appropriations for the army, navy and continental subsidies. As we have seen, there was a substantial element of sleight of hand in Pitt's domestic rhetoric and policy, but in the end Pitt's success lay not so much in manipulation but in persuading a diverse coalition to support a strategy which many of them had initially regarded with extreme scepticism. In this way, Pitt's wartime leadership was sustained by a second virtuous circle: his popularity could be parlayed into taking strategic risks, the success of which increased his popularity. All this could not last for ever, of course, and the Pittite system disintegrated after 1760, when the death of accession of King George III, the outbreak of war with Spain and a whole host of other factors changed the international and domestic framework within which British strategy operated. Pitt himself resigned in October 1761, in protest over the failure to launch a pre-emptive war with Spain. All this can be taken as evidence that the political leadership he had shown until then was bound to hit the buffers sooner or later, but one can equally argue that it was nothing short of miraculous that Pitt had been able to square so many strategic and domestic circles for so long.

Frank-Lothar Kroll

Reformstreben, Machtpolitik und Staatsführung im Krieg bei William Pitt dem Jüngeren

I.

William Pitt der Jüngere gehört in mehrfacher Hinsicht zu den historischen Ausnahmeexistenzen der neueren europäischen Geschichte. Geboren als zweiter Sohn des Earl of Chatham, William Pitt des Älteren, im Jahr 1759, dem *annus mirabilis* seines Vaters – einem Glücksjahr für England, das auf dem Höhepunkt erfolgreicher Kriegführung gegen den französischen Rivalen um die koloniale Vorherrschaft in Nordamerika stand –, brillierte der jüngere Pitt zunächst durch seine erstaunliche Frühreife.[1] Verstandesschärfe, Wissensdrang und Lernfreude zeichneten ihn schon in der Kindheit aus, so daß manche meinen mochten, er sei niemals jung gewesen. Ein kurzes Studium am College *Pembroke Hall* in Cambridge sollte ihn für die juristische Laufbahn qualifizieren und ihm zugleich gediegene Kenntnisse der lateinischen und griechischen Klassiker, der französischen Sprache und der philosophischen Hauptschriften des Jahrhunderts vermitteln. 1781, noch nicht zweiundzwanzigjährig, trat er als Parlamentsabgeordneter in das *House of Commons* ein, dem er bis zu seinem Tod ohne Unterbrechung an-

---

1 Zur Biographie vgl. neben dem zweibändigen „Klassiker“ von *John Holland Rose*, Pitt and the National Revival. London 1911; *ders.*, Pitt and the Great War. London 1911, jetzt die monumentale dreibändige, ganz aus den Quellen erarbeitete und für lange Zeit gültige Darstellung von *John P. W. Ehrman*, The Younger Pitt. The Years of Acclaim. London 1969; *ders.*, The Younger Pitt. The Reluctant Transition. London 1983; *ders.*, The Younger Pitt. The Consuming Struggle. London 1997. Knapper *Robin Reilly*, Pitt the Younger 1759–1806. London 1978; als Essay noch immer anregend *Erich Marcks*, Der jüngere Pitt und seine Zeit (1906). Wiederabgedruckt in: Ders., Männer und Zeiten. Aufsätze und Reden zur neueren Geschichte, Bd. 1. Leipzig 5. Aufl. 1918, S. 97–126; zur Jugendentwicklung speziell *Felix Salomon*, William Pitt, Bd. 1 [mehr nicht erschienen]: Bis zum Ausgang der Friedensperiode (1793). Leipzig 1901, bes. S. 1–38.

gehörte. Im Jahr darauf wurde er britischer Schatzkanzler und übernahm 1783, als kaum Vierundzwanzigjähriger, das Amt des Premierministers immerhin eines der fünf führenden europäischen Länder. Danach leitete er für mehr als zwei Jahrzehnte – mit nur einer kurzzeitigen Unterbrechung – die Geschicke Englands als dessen führender Staatsmann in stark bewegter Zeit. Und er starb 1806, noch nicht siebenundvierzigjährig – im gleichen Lebensjahr stand Otto von Bismarck, als dessen Laufbahn im preußisch-deutschen Staatsdienst überhaupt erst begann.

Man wird nicht fehlgehen, die Pitt'sche Ministerpräsidentschaft zu einer der bedeutendsten der englischen Geschichte zu rechnen. In außenpolitischer Hinsicht dürfte sie wohl die bedeutendste überhaupt gewesen sein. Denn William Pitt der Jüngere schuf nicht nur die Fundamente des zweiten britischen Kolonialreiches, das bis in die ausgehenden 1940er Jahre Bestand haben sollte, nachdem sein König, Georg III., das erste, nordamerikanische Imperium 1782 durch die Intransigenz seiner Überseepolitik leichtfertig verspielt hatte. Der jüngere Pitt agierte darüberhinaus auch ein Jahrzehnt hindurch als wichtigster, zeitweise als der alleinige Gegner Napoleon Bonapartes und schuf dadurch die Voraussetzung für die 1815, ein Jahrzehnt nach seinem frühen Tod, glücklich vollendete Befreiung des Kontinents von der lastenden Hegemonie des korsischen Diktators. Daß damit zugleich die Epoche eines hundertjährigen Krieges mit Frankreich für England zu Gunsten des Inselreiches definitiv ihr Ende fand, verleiht Pitt, zumindest aus britischer Perspektive, den Anspruch auf historische Größe.

Dabei waren Persönlichkeit, Habitus und Charakter des jüngeren Pitt nicht unbedingt dazu angetan, ihn dauerhaft und erfolgreich in einer derart exponierten Stellung verweilen zu lassen. Von Geburt an kränkelnd, schwächlich und fragil, wurde seine prekäre körperliche Konstitution durch einen geradezu unglaublich anmutenden Konsum schweren Portweins zusätzlich untergraben – Pitt pflegte diesen Konsum nicht nur im privaten Rahmen, sondern auch während der zahlreichen Parlamentssitzungen, und niemals wäre es ihm eingefallen, rednerischer Heiserkeit etwa mit einem Glas Wasser zu begegnen. Dürr und hochgewachsen, förmlich und steif, von kühler, zuweilen hochmütig wirkender Zurückhaltung, bestach Pitt dann doch durch seine glänzende Rednergabe, seinen Scharfsinn und seine Schlagfertigkeit, überzeugte durch die Klarheit und Sachlichkeit seiner Argumentation und, nicht zuletzt, durch seine unermüdliche Tatkraft und seinen rastlosen Arbeitseifer. Als die Frau seines hartnäckigsten parlamentarischen Gegenspielers Charles James Fox ihn zum ersten Mal erblickte, bemerkte sie entsetzt: *„Was für ein häßlicher Mensch!“* *„Sprich nicht so“*, entgegnete ihr

Fox, „*wenn Du ihn sähest, während er eine Rede hält, würdest Du ihn wundervoll finden*".[2]

Pitts Amtszeit kann in zwei klar voneinander getrennte Abschnitte eingeteilt werden, die ihren inneren Zusammenhang durch die in ihnen jeweils vorherrschende situationsbedingte Akzentsetzung erhalten. Der erste Abschnitte umfaßte das Friedensjahrzehnt zwischen 1783 und 1793, und er war geprägt vom Vorgang des inneren Landesausbaus, von zahlreichen Reformvorhaben, die allerdings nur in einigen Fällen realisiert werden konnten, besonders aber von der institutionellen Konsolidierung der Position des britischen Premierministers als solcher und von der Wiederherstellung geordneter staatlicher Finanzverhältnisse. All das waren unverzichtbare Voraussetzungen für jene erstaunlichen Erfolge, die Pitt dann in der zweiten Phase seiner Amtsführung für sich verbuchen konnte. Diese zweite Phase umfaßte die Jahre von 1793 bis 1801 beziehungsweise von 1804 bis 1806. Sie stand ganz im Zeichen der Außenpolitik, und sie war geprägt vom ununterbrochenen militärischen Engagement britischer Truppen in den verschiedensten Erdteilen, keineswegs nur in Europa. Daher muß ein Beitrag, der dem Problem staatsmännischen Wirkens in Kriegs- und Krisenzeiten gewidmet ist und Pitts *reforming leadership in war* zu rekonstruieren strebt, seinen Argumentationsgang mit einer kurzen Inblicknahme jener zunächst im Friedenjahrzehnt bis 1793 von ihm selbst geschaffenen inneren Voraussetzungen eröffnen, ohne die sein späteres Wirken als Kriegspremier undenkbar gewesen wäre.

II.

Das langfristig wohl wichtigste innenpolitische Resultat des ersten Jahrzehnts der Pitt'schen Ministerpräsidentschaft bestand in der definitiven Verankerung der Position des Premierministers gegenüber den Herrschaftsansprüchen der Krone. Man weiß, daß seit den Zeiten Robert Walpoles, also in den 1720er und 1730er Jahren, das *House of Commons* eine bedeutsame Machtsteigerung erfahren hatte, und daß sich damals ansatzweise bereits jene Regierungsform auszubilden begonnen hatte, die später mit dem Begriff *cabinet system* bezeichnet worden ist. Dieses „System" verband das Prinzip der Volkswahl des Ersten Ministers mit dessen Verantwortlichkeit gegenüber der Parlamentsmehrheit, bei gleichzeitiger relativer Unabhängigkeit von unmittelbaren Beeinflussungsversuchen der Krone.[3] In den beiden

---

2 Zitiert nach *Erich Eyck*, Die Pitts und die Fox' – Väter und Söhne. Zwei Paar verschlungener Lebensläufe. Erlenbach-Zürich 1946, S. 397.

3 Dazu umfassend, wenngleich nur noch mit Einschränkungen benutzbar *Hermann Savelkouls*, Das englische Kabinettsystem. München/Berlin 1937, bes. S. 26 ff., 158 ff.,

ersten Jahrzehnten der Regierungszeit Georgs III., zwischen 1760 und 1782, hatte der König dieses „System" dann aber schrittweise durch ein persönliches Regiment zu ersetzen gewußt. Es war sein Ziel gewesen, den Einfluß der Parteien, vor allem der *Whigs*, dauerhaft auszuschalten und stattdessen die Mehrheitsverhältnisse im Parlament mit einer eigenen Gefolgschaft, *the King's friends*, zu manipulieren.[4] Der Wille der Krone sollte hinfort durch Bestechungen und Begünstigungen innerhalb wie außerhalb des Parlaments als vorherrschender Faktor im Verfassungsleben installiert werden. Georg III. war mit diesem Konzept solange erfolgreich gewesen, bis der unglückliche Kriegsausgang in Nordamerika sein System persönlicher Königsherrschaft 1782 zum Einsturz gebracht hatte. Nach dem kurzen Zwischenspiel der (zweiten) Regierung Rockingham – die eine Politik der Versöhnung mit den Vereinigten Staaten einleitete[5] – und der Fox-North-Koalition – die erste englische Regierung, die der Krone von einer Unterhausmehrheit aufgezwungen wurde[6] –, war Pitt zwar noch vom König selbst berufen worden,[7] und er hat Georg III. gegenüber, trotz dessen Engstirnigkeit und Kleinlichkeit, stets strikte Loyalität bekundet.[8] Doch er hat zugleich dafür gesorgt, daß die gezielte Einflußnahme des Königs auf ausgewählte Parlamentarier unterblieb, daß der Monarch einzelne Ressortminister innerhalb der Regierung nicht mehr gegeneinander ausspielen konnte, und daß sich

---

170 ff., 213 ff., 298 ff.; ferner als historischer Gesamtüberblick *John P. Mackintosh*, The British Cabinet. London 1962, sowie mit speziellem Bezug auf die Entwicklung unter dem jüngeren Pitt *Arthur Aspinall*, The Cabinet Council, 1783–1835. In: Proceedings of the British Academy 68 (1952), S. 145–252.

4 Zur diesbezüglichen Politik Georgs III. noch immer unentbehrlich *Richard Pares*, King George III and the Politicians. Oxford 1953, S. 61 ff., 143 ff., sowie als beste Biographie des umstrittenen Monarchen *John Brooke*, King George III. With a Foreword by H. R. H. The Prince of Wales. London 1972, bes. S. 217–259; zur Regierungsweise des Königs speziell *Ian R. Christie*, The Cabinet in the Reign of George III, to 1790. In: Ders., Myth and Reality in Late-Eighteenth-Century British Politics and Other Papers. London 1970, S. 55–108; aus „whiggistischer" Perspektive *Herbert Butterfield*, George III and the Constitution. In: History, N. S. 43 (1958), S. 14–33.

5 Vgl. *Frank O'Gorman*, The Rise of Party in England. The Rockingham-Whigs, 1760–82. London 1975.

6 Vgl. detailliert *John Cannon*, The Fox-North-Coalition. Crisis of the Constitution 1772–84. Cambridge 1969, bes. S. 228–243; ferner *Robert Blake*, 1783–1902. In: David Butler (Hrsg.), Coalitions in British Politics. London 1978, S. 1–24.

7 Zu Pitts Anfängen *Paul Kelly*, British politics, 1783–4. The Emergence and triumph of the Younger Pitt's Administration. In: Bulletin of the Institute of Historical Research 54 (1981), S. 62–78.

8 Dazu – neben der Quellenedition von *John Holland Rose*, The King and Pitt. In: Ders., Pitt and Napoleon. Essays and Letters. London 1912, S. 201–248 – noch immer sehr lesenswert *Donald G. Barnes*, George III and William Pitt 1783–1806. A new Interpretation based upon a Study of their unpublished Correspondence (1939). Neudruck New York 1965.

keine neuerliche Günstlingswirtschaft der Krone etablierte.[9] Seit Pitt dem Jüngeren, und durch ihn, war der britische Premierminister tatsächlich – als Haupt eines mit ihm solidarischen Ministeriums – der erste und wirklich Leiter der gesamten Regierungsgeschäfte. Er war (und blieb) als solcher nicht mehr nur dem Monarchen gegenüber verantwortlich, und er agierte erst recht nicht länger als dessen bloßes Werkzeug. Vielmehr präsidierte er seither dem Kabinett als einer primär von den Mehrheitsverhältnissen im *House of Commons* abhängigen Einrichtung.[10]

Die so gewonnene Machtposition ermöglichte Pitt über zwei Jahrzehnte hinweg ein weitgehend problemloses Regieren. Er handelte im Einklang mit der Krone, die er stützte, deren machtpolitische Begehrlichkeiten er jedoch zugleich in Grenzen hielt, und für die er zunehmend unentbehrlich wurde. Zugleich wußte er sich in Übereinstimmung mit dem Parlament, in dessen Reihen er, gestützt auf entsprechende Wahlergebnisse, zumeist mit klaren Mehrheiten rechnen konnte, und das er durch seine glänzende Rhetorik[11] in vielen, freilich nicht in allen strittigen Fällen von der Richtigkeit des von ihm vertretenen Kurses zu überzeugen vermochte.

Es war vor allem die von vielen zeitgenössischen Beobachtern als vordringlich empfundene Reform des Wahlrechts, die Pitt zu Beginn seiner Ministerpräsidentschaft stark beschäftigte,[12] von ihm jedoch damals nicht erfolgreich auf den Weg gebracht werden konnte. Ein diesbezüglich einge-

---

9 Dazu maßgeblich bereits *Archibald S. Foord*, The Waning of "The Influence of the Crown". In: English Historical Review 62 (1947), S. 484–507, sowie zum langfristigen Wandel der Machtrelationen zwischen Parlament und Krone *Betty Kemp*, King and Commons, 1660–1832. London 1957.

10 Zur Parlamentsgeschichte in der Ära des jüngeren Pitt vgl. – neben dem umfassenden und gelehrten Großunternehmen von *Sir Lewis B. Namier* und *John Brooke*, The House of Commons 1754–1790, Bde. 1–3. London 1964 (mit Betonung der noch weitgehend parteiungebundenen Handlungsweise der Parlamentsmitglieder), und dem ebenso gelehrten Fortsetzungswerk von *Roland Thorne*, The House of Commons 1790–1820, Bde. 1–5. London 1986 – vor allem *Peter D. G. Thomas*, The House of Commons in the Eighteenth Century. Oxford 1971, dort speziell (S. 187 ff.) über parlamentarische Geschäftsformen und Verfahrensweisen zu Pitts Amtsbeginn bis 1784. – Pitt hatte seinen Platz am 23. Januar 1781 als Abgeordneter von Appleby im Unterhaus eingenommen, und er blieb dessen Mitglied für die Dauer von exakt fünfundzwanzig Jahren, bis zum 23. Januar 1806, dem letzten Tag seines Lebens.

11 Dazu *John Holland Rose*, The Oratory of Pitt. In: Ders., Pitt and Napoleon (wie Anm. 8), S. 1–19.

12 Darüber im Zusammenhang *John Cannon*, Parliamentary Reform 1640–1832. Cambridge 1973, bes. S. 86 ff., 126–139; spezieller *Frank O'Gorman*, The Unreformed Electorate of Hanoverian England. The Mid-Eighteenth Century to the Reform Act of 1832. In: Social History 11 (1986), S. 33–52, sowie *ders.*, Voters, Patrons, and Parties. The Unreformed Electoral System of Hanoverian England 1734–1832. Oxford 1989, bes. S. 295 f.

brachter Gesetzesentwurf fiel 1785 in der parlamentarischen Abstimmung glatt durch.[13] Die Annahme der Pitt'schen Vorlage hätte zur Beseitigung der „abgestorbenen" Wahlkreise mit zum Teil nur ganz wenigen Wahlberechtigten geführt – jener berüchtigten *rotten boroughs,* in denen es praktisch möglich war, sämtliche Stimmen für eine Wahl zum Parlamentsmitglied zu kaufen. Bei den Unterhauswahlen von 1784, den ersten in Pitts Amtszeit als Premierminister, konnten einstmals bevölkerungsreiche, doch mittlerweile verödete Orte wie Old Sarum (mit elf Wählern),[14] Newtown (mit 23 Wählern) oder Dunwich (mit 32 Wählern) jeweils zwei Abgeordnete in das *House of Commons* entsenden, während etwa eine Großstadt wie Manchester, deren Einwohnerzahl seit Beginn der Industrialisierung zur Jahrhundertmitte stark gewachsen war, keinen einzigen Abgeordneten als Parlamentsvertreter in London aufzuweisen vermochte. Die von Pitt erstrebte behutsame Anpassung der Wahlkreisgrößen an die veränderten Bevölkerungsverhältnisse durch Ausdehnung des Wahlrechts auf die einwohnerstarken Grafschaften und auf einige Städte wurde bekanntlich erst 1832 erreicht.[15] Pitt selbst ist in seinen späteren Jahren nicht wieder auf die Neuregelung der Wahlrechtsfrage zurückgekommen, wofür wohl nicht zuletzt die brisante außenpolitische Lage Großbritanniens mitverantwortlich gewesen sein dürfte.

Die zweite wichtige innenpolitische Voraussetzung für Pitts dauerhaft betriebenes europa- und weltpolitisches Engagement bildete zweifellos seine überaus erfolgreiche Finanz-und Wirtschaftspolitik. 1783, zum Zeitpunkt seines Amtsantritts, befanden sich die britischen Staatsfinanzen in einer heillos zerrütteten Verfassung. England stand, besonders infolge der hohen Kosten für die Kriegführung in Nordamerika, am Rand des Staatsbankrotts.[16]

---

13 Für den Zusammenhang *Ian R. Christie*, Wilkes, Wyvill and Reform. The Parliamentary Reform Movement in British Politics 1760–1785. London 1962.

14 Diesem Wahlkreis verdankte Pitts Vater, William Pitt der Ältere, bekanntlich seinen Einzug ins Parlament 1735.

15 Den Entwicklungsgang von der Amtszeit des jüngeren Pitt bis zum *Great Reform Act* 1832 skizziert – neben den noch immer mustergültigen „Klassikern" von *Edward* und *Annie G. Porritt*, The Unreformed House of Commons. Parliamentary Representation before 1832 (1903–1909), Bde. 1–2. Nachdruck London 1963, und *George S. Veitch*, The Genesis of Parliamentary Reform (1913). Nachdruck London 1965 – sehr instruktiv *Michael G. Brock*, The Great Reform Act. London 1973; einen noch umfassenderen Argumentationsstandpunkt bezieht *William D. Rubinstein*, The End of "Old Corruption" 1780–1860. In: Past and Present 101 (1983), S. 55–86.

16 Dazu *John E. D. Binney*, British Public Finance and Administration, 1774–1792. Oxford 1958, sowie *John Brewer*, The Sinews of Power. War, Money and the English State 1688–1783. London 1989; ferner die ältere, weiterhin wichtige Studie von *David L. Keir*, Economical Reform, 1779–1787. In: Law Quarterly Review 50 (1934), S. 368–385.

Pitt hat die maroden britischen Staatsfinanzen in erstaunlich rascher Zeit zu sanieren und danach dauerhaft in Ordnung zu halten vermocht, was in der Rückschau auch deshalb einigermaßen erstaunlich anmuten mag, weil er in seiner privaten Lebensführung alles andere als ein sparsamer Haushalter, vielmehr ein geradezu exorbitanter Verschwender gewesen ist. Sorglos und nachlässig in allen persönlichen Geldangelegenheiten, war er bei seinem Tod hochverschuldet – die von den Gläubigern eingeforderte sagenhafte Summe von mehr als 40.000 Pfund hat der britische Staat, gleichsam als Dankesschuld gegenüber seinem großen Politiker, diskret beglichen.

Pitts Rezept zur Beilegung der staatlichen Finanzkrise bestand aus einem Bündel ebenso mutiger wie unkonventioneller Maßnahmen. Mochten die Einrichtung eines Tilgungsfonds zur Verringerung der Staatsschulden, die Herabsetzung des Warenzolls zur Reduzierung des Schmuggels, die Vereinfachung des Systems der Abgaben und die Erhöhung der direkten Steuern für Luxusgüter noch zum einigermaßen gewohnten Repertoire britischer Steuer- und Finanzpolitik im 18. Jahrhundert zählen[17] – die 1789 von Pitt durchgesetzte Einführung einer progressiven Einkommenssteuer war ein für britische Verhältnisse höchst ungewohntes Instrument.[18] Sie unterwarf den Verdienst eines jeden Bürgers – falls er denn einen solchen vorweisen konnte – in abgestuften Graden der Steuerpflicht, deren Höhe sich nach der jeweiligen Wohlhabenheit des Besteuerten richtete. Trotz vereinzelt artikulierter Gegnerschaft haben die Briten diese Belastung, wie von Pitt vorgesehen, bis zum Ende des Krieges mehrheitlich akzeptiert.[19] Man wird nicht fehlgehen, diese kollektive Bereitschaft zur Übernahme zusätzlicher finanzieller Opfer als eine wichtige Vorbedingung für die weitgehend ungebrochene Leistungskraft Großbritanniens in den weltpolitischen Auseinander-

---

17 Dazu – neben dem weiterhin unverzichtbaren monumentalen Standardwerk von *Stephen Dowell*, A History of Taxation and Taxes in England from the Earliest Times to the Present Day. Bde. 1–4. London 1884 – *John V. Beckett* und *Michael Turner*, Taxation and Economic Growth in Eighteenth Century England. In: Economic History Review, 2nd series 43 (1990), S. 377–403.

18 Dazu neben der älteren Studie von *William Kennedy*, English Taxation 1640–1799 (1913). Nachdruck London 1964, allgemein *Basil E. V. Sabine*, A History of the Income Tax. London 1966.

19 Zur Entwicklung bis 1815 *Arthur Hope-Jones*, Income Tax in the Napoleonic Wars. Cambridge 1939, sowie in größerem Rahmen *Patrick O'Brien*, The Political Economy of British Taxation, 1660–1815. In: Economic History Review, 2nd series 41 (1988), S. 1–32: kritisch gegenüber Pitts steuerpolitischen Unternehmungen steht *John R. Breihan*, William Pitt and the Commission on Fees, 1785–1801. In: Historical Journal 27 (1984), S. 59–81.

setzungen der kommenden Jahre zu werten. Denn durch Anleihen allein[20] waren die rasant wachsenden Bedürfnisse des Krieges nicht mehr einzulösen.

Noch weitaus zukunftweisender als seine finanzpolitischen Maßnahmen – selbst ein sonst so kritisch urteilender Beobachter wie Erich Eyck nannte Pitt anerkennend den „größte[n] Finanzpolitiker seiner Zeit“[21] – war Pitts wirtschafts- und handelspolitisches Engagement. Hier offenbarte sich der politisch so konservativ eingestellte Führer und Erneuerer der *Tory*-Partei[22] als unermüdlicher Vorkämpfer des liberalen Freihandelsgedankens und als gelehriger Schüler von Adam Smith, dessen 1776 erschienenes Werk *An Inquiry into the Nature and Causes of the Wealth of Nations* er eingehend studiert und rezipiert hatte.[23] Darüberhinaus kann er, mit gewissen Einschränkungen, auch als ein früher Vorläufer von Richard Cobden gelten. Der von Cobden 1839 begründeten *Anti-Corn Law League* gelang es bekanntlich 1846, die Abschaffung aller englischen Zölle und Handelsbeschränkungen durchzusetzen. Zwar konnte Pitt derart weitgesteckte Ziele in den 1780er Jahren noch nicht realisieren. Doch den Weg vom staatlichen Merkantilismus zur bürgerlichen Wirtschaftsfreiheit, den – mit zeitlicher Verzögerung – später auch alle anderen europäischen Länder beschreiten sollten, hat der jüngere Pitt als erster britischer Staatsmann eingeschlagen. Der in diesem Sinn wegweisende Handelsvertrag Englands mit Frankreich vom 27. September 1786 legte den langjährigen Zollkrieg zwischen den beiden Konkurrenten bei, ermöglichte einen freieren Warenverkehr mit bilateralem Güteraustausch und kam mit solchen Bestimmungen vor allem dem Export britischer Industrieprodukte zu Gute.[24] Eine ähnliche Vereinbarung

---

20 Aufschlußreiche Analysen zu Umfang und Ausmaß der britischen Staatsverschuldung im Zeitalter des Merkantilismus bietet *Alice C. Carter*, The English Public Debt in the 18th Century. London 1968.

21 *E. Eyck*, Die Pitts und die Fox' (wie Anm. 2), S. 393.

22 Dazu im Zusammenhang *Keith G. Feiling*, The Second Tory Party, 1714–1832 (1938). Nachdruck London 1959.

23 In seiner Budgetrede von 1792 bekannte sich Pitt offen als Anhänger der Lehren von Adam Smith; vgl. *F. Salomon*, William Pitt (wie Anm. 1), S. 112–206; nach Auffassung Walter Bagehots, des maßgeblichen Interpreten englischer Verfassungszustände im 19. Jahrhundert, war der jüngere Pitt „the first Englishman who read, understood and valued the *Wealth of Nations*“; zitiert nach *Walter Bagehot*, William Pitt. In: Ders., Collected Works. Hrsg. von Norman St. John-Stevas. Bde. 1–15. Hier Bd. 3. London 1968, S. 123–155, S. 130.

24 Vgl. – neben den älteren Studien von *John Holland Rose*, The Franco-British Commercial Treaty 1786. In: English Historical Review 23 (1908), S. 709–724, und *William Otto Henderson*, The Anglo-French Commercial Treaty of 1786. In: Economic History Review, 2nd series 10 (1957/58), S. 104–122 – für die außenpolitischen Dimensionen des Vertragswerkes speziell *Mary M. Donaghay*, The Vicious Circle. The

hatte Pitt bereits 1785 für den Handelsverkehr zwischen England und Irland geplant, um der irischen Wirtschaft, die sich in einem noch sehr bescheidenen Ausmaß zu regen begann, freien Zugang zu britischen Märkten und kolonialen Absatzgebieten zu verschaffen.[25] Doch die übertriebenen Befürchtungen Londoner Handelskreise vor einer Überschwemmung des Inlandsmarktes mit billigen irischen Erzeugnissen brachten eine entsprechende Gesetzesvorlage zu Fall.

III.

Die zweite Phase der Pitt'schen Amtsführung stand ganz im Zeichen außenpolitischer Verwicklungen, und das hieß konkret: kriegerischen Engagements in Europa und Übersee. Das Zentrum dieser Verwicklungen und dieses Engagements, zugleich der Ausgangs- und stete Bezugspunkt für Pitts gesamtes weiteres staatsmännisches Handeln war Frankreich, waren die seit 1789 von Paris ausgehenden revolutionären Erschütterungen. Pitt hatte diese Erschütterungen, wie viele seiner Landsleute, zunächst ausdrücklich begrüßt, denn er erblickte in einem durch innere Kämpfe gelähmten bourbonischen Königtum eine willkommene Schwächung des traditionellen politischen und ökonomischen Rivalen England. Die Krise der Monarchie in Frankreich wurde in London mit unverholener Genugtuung wahrgenommen, man sah in ihr einen Prozeß der Selbstzerstörung und Selbstauflösung, dessen rasanter Fortschritt der britischen Politik ungeahnte Möglichkeiten zur weltweiten Machtausdehnung zu eröffnen schien. Lange Zeit hat Pitt denn auch keineswegs daran gedacht, etwa zugunsten der bedrohten französischen Monarchie zu intervenieren.[26] Stattdessen bevorzugte er eine Politik des Friedens, der Neutralität und der Nichteinmischung in die inneren Angelegenheiten Frankreichs.

Mit dieser Haltung unterschied er sich grundlegend von der Position seines Bundesgenossen Edmund Burke, des wohl prominentesten publizistischen Kämpfers gegen die Prinzipien der Französischen Revolu-

---

Anglo-French Commercial Treaty of 1786 and the Dutch Crisis of 1787. In: Proceedings of the Consortium on Revolutionary Europe 19 (1989), S. 447–456; ferner *John P. W. Ehrman*, The British Government and Commercial Negotiations with Europe, 1783–1793. Cambridge 1962.

25 Dazu umfassend *Lewis M. Cullen*, Anglo-Irish Trade, 1660–1800. Manchester 1968; ferner die Spezialstudien von *Paul Kelly*, British and Irish Politics in 1785. In: English Historical Review 90 (1975) S. 536–563, und *James Kelly*, The Anglo-French Commercial Treaty of 1786. The Irish Dimension. In: Eighteenth-Century Ireland 4 (1989), S. 93–111.

26 Pitts Einstellung zu Frankreich in den Jahren bis 1793 resümiert die ältere, aus französischen und britischen Archivquellen gearbeitete Studie von *John T. Stoker*, William Pitt et la Révolution Française (1789–1793). Paris 1935.

tion.[27] Burkes politische Karriere hatte bekanntlich im Kreis jener liberal-aristokratisch gesinnten *Whigs* begonnen, die in den 1780er Jahren einen „antidespotischen Konstitutionalismus" verfochten hatten, welcher eine Einschränkung monarchischer Machtfülle mittels parlamentarischer Kontrollorgane favorisiert hatte. 1791 war es dann zum offenen Bruch mit dem Führer der *Whigs* im Unterhaus, Charles James Fox, gekommen, weil Burke den von Fox vertretenen revolutionsfreundlichen und pro-französischen Kurs nicht länger mittragen mochte.[28] Burke hatte daraufhin die zusehends zerfallende *Whig*-Partei, gemeinsam mit einigen politischen Freunden, verlassen[29] und warb seither mit glühendem Eifer unermüdlich für einen Kreuzzug gegen die Pariser Republik zerlumpter Sansculotten, die er als Todfeindin der gesamten zivilisierten Welt identifizierte.[30]

Diesen fanatischen Aufrufen Burkes ist Pitt nur sehr zögernd gefolgt.[31] Weder der Sturz der französischen Monarchie im August 1792 noch die

---

27 Aus der reichhaltigen Literatur vgl. vor allem *Dietrich Hilger*, Edmund Burke und seine Kritik der Französischen Revolution. Stuttgart 1960; *Wolfgang J. Mommsen*, Edmund Burke und die Französische Revolution. In: Kurt Kluxen und Wolfgang J. Mommsen (Hrsg.), Politik, Ideologie und nationalstaatliche Ordnung. Studien zur Geschichte des 19. und 20. Jahrhunderts. Festschrift für Theodor Schieder. München/Wien 1968, S. 39–66; *Albert Goodwin*, The political genesis of Burke's "Reflections on the Revolution in France". In: Bulletin of The John Rylands Library 50 (1968), S. 336–364; *Michael Freeman*, Edmund Burke and the Critique of Political Radicalism. Oxford 1980. Für den Zusammenhang wichtig die ältere Studie von *Stephan Skalweit*, Edmund Burke und Frankreich. Köln/Opladen 1956, bes. S. 29 ff.

28 Dazu detailliert *Frank O'Gorman*, The Whig party and the French Revolution. London 1967, bes. S. 175–180; *Leslie G. Mitchell*, Charles James Fox and the Disintegration of the Whig Party, 1782–1794. Oxford 1971, bes. S. 153–193; vgl. ferner bereits *Herbert Butterfield*, Charles James Fox and the Whig Opposition in 1792. In: Cambridge Historical Journal 9 (1947/49), S. 293–330; die Fortentwicklung der sich auch rhetorisch immer stärker radikalisierenden Positionen von Fox skizzieren *Loren Reid*, Charles James Fox. A Man for the People. London/Harlow 1969, S. 251 ff., 308 ff., 361–371; *John W. Derry*, Charles James Fox. London 1972, sowie zuletzt *Leslie G. Mitchell*, Charles James Fox. Oxford 1992.

29 Zur ideologischen „Vorgeschichte" der Parteispaltung vgl. in diesem Zusammenhang *Brian W. Hill*, Fox and Burke. The Whig party and the Question of Principles, 1784–1789. In: English Historical Review 89 (1974), S. 1–24.

30 Vgl. The Works and Correspondences of the Right Honourable *Edmund Burke*. A new Edition. Hrsg. von Charles W. Wentworth-Fitzwilliam und Richard Bourke. Bde. I–VIII, Bd. IV. London 1852, hier besonders die im Verlauf des Jahres 1791 erschienenen Schriften *An Appeal from the New to the Old Whigs* (S. 393–490), *A Letter to a Member of the National Assembly* (S. 357–392), sowie *Thoughts an French Affairs* (S. 551–591). Vor allem mit der letztgenannten Abhandlung, der reifsten und gedanklich geschlossensten seiner Revolutionsschriften, wollte Burke die europäischen Kabinette zu einem aktiven militärischen Vorgehen gegen die französischen Revolutionäre veranlassen.

31 Zur Entwicklung der Pitt'schen Frankreichpolitik im Schatten der Revolution bis zum Kriegsausbruch 1793 detailliert *J. H. Rose*, Pitt and the Great War (wie Anm. 1),

abstoßenden Mordorgien der jakobinischen Terroristen in den Pariser Gefängnissen – die berüchtigten Septembermassaker von 1792 –, weder die schrankenlose Herrschaft der Guillotine noch selbst die in ganz England mit Ekel und Abscheu aufgenommenen Begleitumstände der Ermordung Ludwigs XVI. im Januar 1793,[32] konnten ihn dazu veranlassen, sich dem seit April 1792 geführten Interventionskrieg Österreichs und Preußens gegen das revolutionäre Frankreich anzuschließen. Erst als die Revolutionstruppen nach der Schlacht von Jemappes im November 1792 die Österreichischen Niederlande eroberten, Holland bedrohten und sich zur Besetzung des linken Rheinufers anschickten, erst dann, Anfang Februar 1793, und nicht früher,[33] führte auch Pitt sein Land in die große militärische Auseinandersetzung zwischen den Mächten Alteuropas und der Revolution.

Die Wahl gerade dieses Zeitpunktes erscheint für die politische Weltsicht des jüngeren Pitt außerordentlich bezeichnend. Maßgeblich war für ihn nicht, wie für seinen konservativen Gesinnungsfreund Burke, der „Krieg gegen ein System, das seinem Wesen nach allen anderen Regierungen feindlich gesinnt ist und zum Frieden oder zum Krieg schreitet, je nachdem Friede oder Krieg am besten geeignet sind, diese zu stürzen“[34]. Im Gegenteil: „Es ist kindisch und ein Zeichen schwachen Verstandes, behaupten zu wollen, daß eine Nation unveränderlich die Feindin einer anderen sein könne; weder die Erfahrung der Nationen noch die Geschichte der Menschheit liefert hierfür einen Beweis“.[35] In diesen Worten Burkes und Pitts kommen auf geradezu idealtypische Weise zwei unterschiedliche Möglichkeiten konservativer Politikauffassung zum Ausdruck, wie sie dann verstärkt im 19. Jahrhundert, und hier keineswegs nur in England,[36] sondern beispielsweise auch in Preu-

S. 72–117; *J. P. W. Ehrman*, The Reluctant Transition (wie Anm. 1), S. 233–257, sowie *Jeremy Black*, Anglo-French relations in the age of the French Revolution 1787–1793. In: Francia 15 (1987), S. 407–433.

32 Dazu direkt *Albert Mathiez*, Danton, Talon, Pitt et la mort de Louis XVI. In: Annales Révolutionnaires 8 (1916), S. 367–376.

33 Zu den Umständen des Kriegseintritts vgl. bereits die Miniatur von *John Holland Rose*, Pitt and the Campaign of 1793 in Flanders. In: English Historical Review 24 (1909), S. 744–749; zuletzt *Philip Schofield*, British politicians and French arms. The ideological war of 1793–1795. In: History 77 (1992), S. 183–201.

34 *Edmund Burke*, Letters on a Regicide Peace (1796); zitiert nach *John Holland Rose*, Der jüngere Pitt (1938). 2. Aufl. München 1948, S. 150.

35 Statement *Pitts* im *House of Commons* anläßlich der Unterzeichnung des britisch-französischen Handelsvertrages 1786; zitiert nach *Felix Salomon*, Der britische Imperialismus. Ein geschichtlicher Überblick über den Werdegang des britischen Reiches vom Mittelalter bis zur Gegenwart. Berlin 1916, S. 114.

36 Für die dortige Tradition aufschußreich *James J. Sack*, The memory of Burke and the memory of Pitt. English conservatism confronts its past, 1806–1829. In: Historical Journal 30 (1987), S. 623–640.

ßen, etwa im Blick auf die beiden konservativen Antipoden Ernst Ludwig von Gerlach und Otto von Bismarck, nachzuweisen sind: die konservativ-ideenpolitische Haltung, vertreten durch Burke (und später durch Gerlach) einerseits, und die konservativ-realpolitische Haltung, vertreten durch Pitt (und später durch Bismarck) andererseits.

Tatsächlich war Pitts 1793 eröffneter Krieg gegen Frankreich durchaus kein gegenrevolutionär motivierter Prinzipienkampf gegen das Denken und Handeln der Königsmörder in Paris, so wie Burke sich diesen gewünscht hatte. Burke hat denn auch mit kritischen Worten gegenüber der Frankreichpolitik der Regierung Pitt nicht gespart.[37] Seiner Auffassung zufolge krankte Pitts gesamte Einstellung an einer fatalen Unterschätzung gerade der ideologischen Herausforderung, mit der die revolutionäre Bewegung alle Staaten und Nationen Europas gleichermaßen bedrohe. In diesem Sinn galt Burke auch die Unterstützung der anti-revolutionären royalistischen Volkserhebung innerhalb Frankreichs, die im Westen des Landes, besonders in der Bretagne und in der Vendée, als „Chouannerie" seit 1793 einen äußerst aktiven Guerillafeldzug gegen die Pariser Machthaber führte,[38] als ein Gebot der politischen Vernunft. Nach dem Eintritt der konservativen *Whigs* um den Duke of Portland, den Earl Fitzwilliam und William Windham ins Kabinett – alle drei waren enge politische Freunde Edmund Burkes und hatten sich 1793/94 wegen des immer radikaleren und revolutionsfreundlichen Kurses ihres Parteiführers Charles James Fox von der *Whig*-Fraktion getrennt[39] – hat Pitt dann zwar seit Juli 1794 im Vollzug entsprechender Koalitionsverabredungen dem Drängen der „Burkeaner" nach einem *ideological war* gegen die Revolution insofern nachgegeben, als er Ende Juni 1795 britische Truppen im Verbund mit Soldaten der royalistischen Emigration an der westfranzösischen Küste landen ließ.[40] Das Expeditionskorps sollte von hier aus die mehrheitlich königstreuen Kräfte im Land zum „Marsch auf Paris" sammeln, dort das verhaßte Direktorium stürzen und die bourboni-

---

37 Grundlegend *Michael Wagner*, England und die französische Gegenrevolution 1789–1802. München 1994, S. 65–71, 94 ff., 126–135, 151 ff., 280–289; ferner die ältere Detailstudie von *John Holland Rose*, Burke, Windham, and Pitt. In: English Historical Review 27 (1912), S. 700–716; 28 (1913), S. 86–105.

38 Dazu maßgeblich *Donald M. G. Sutherland*, The Chouans. The social origins of popular counterrevolution in Upper Britanny, 1770–1796. Oxford 1982.

39 Zu den Einzelheiten *Frank O'Gorman*, The Emergence of the British Two-Party System, 1760–1832. London 1982, sowie *Brian W. Hill*, British Parliamentary Parties 1742–1832. From the Fall of Walpole to the First Reform Act. London 1985, S. 163–177.

40 Vgl. eingehend *M. Wagner*, England und die französische Gegenrevolution (wie Anm. 37), S. 163 ff., sowie die Spezialstudie von *Harvey Mitchell*, The underground war against revolutionary France. The missions of William Wickham 1794–1800. Oxford 1965.

sche Monarchie restaurieren. Doch das schlecht vorbereitete Unternehmen scheiterte kläglich, und überhaupt hat Pitt die partielle Zusammenarbeit mit den Kräften der Gegenrevolution, anders als Burke, niemals zum Selbstzweck erhoben, sondern in ihr stets nur ein Instrument zur Eindämmung revolutionären Expansionsstrebens und zur Förderung britischer Interessen in Europa und Übersee – auf Kosten Frankreichs – erblickt.

Eine solche Haltung stand in vollkommenem Einklang mit den Traditionen englischer Außenpolitik, deren Gesamtrichtung seit den 1690er Jahren durch das Vorherrschen einer deutlich anti-französischen Orientierung geprägt war.[41] Damals hatte das Macht- und Eroberungsstrebens Ludwigs XIV. das europäische Staatensystem mit der akuten Gefahr einer französischen Hegemonialstellung konfrontiert, und schon damals hatte sich das Inselreich mit kriegerischer Entschlossenheit als Anwalt des europäischen Gleichgewichts gegen diese hegemonialen Gelüste Frankreichs zu protegieren gewußt.[42] Nach Formierung der *Großen Allianz* 1689 war dem *roi soleil* 1697 im Frieden von Rijswijk erstmals Einhalt geboten worden, seit 1701 hatte eine erneute Koalition fast aller europäischen Mächte unter Führung Englands die bourbonischen Erbansprüche auf Spanien bestritten und Frankreich nach langem Ringen schließlich im Frieden von Utrecht 1713 zum Verzicht auf seinen kriegerischen Expansionskurs gezwungen. Es war aus dieser weitgespannten Perspektive kein Zufall, daß Pitt gerade 1793 durch den Versuch einer abermaligen Festsetzung der Franzosen an der belgisch-niederländischen Kanalküste zum militärischen Losschlagen bewogen worden ist. Denn an diesem geostrategisch so überaus sensiblen Punkt britischer Sicherheitsinteressen konnte der Zugriff einer starken feindlichen Macht am allerwenigsten geduldet werden. Das war schon

41 Dazu sehr kenntnisreich *Jeremy Black*, Natural and necessary enemies. Anglo-French relations in the eighteenth century. London 1986; allgemeiner *ders.*, A system of ambition? British foreign policy, 1660–1793. London/New York 1991; zur „Vorgeschichte" bis 1789, unter besonderer Berücksichtigung der geistig-kulturellen, aber auch der wirtschafts- und handelspolitischen Wechselbeziehungen *Derek Jarrett*, The Begetters of Revolution. England's involvement with France, 1759–1789. London 1973.

42 Dazu vorzüglich *Kurt Kluxen*, Zur Balanceidee im 18. Jahrhundert (1978). Wiederabgedruckt in: Ders., England in Europa. Studien zur britischen Geschichte und zur politischen Ideengeschichte der Neuzeit. Hrsg. von Frank-Lothar Kroll. Berlin 2003, S. 106–121; ferner die umfassende Erörterung der Thematik im Zusammenhang der Politik der „großen Mächte", vor allem Deutschlands, Frankreichs und Englands (S. 131–211), und unter ausdrücklicher Berücksichtigung der Pitt'schen Gleichgewichtskonzeption als „Ordnungsvorstellung der Politik schlechthin" (S. 169) bei *Eberhard von Vietsch*, Das europäische Gleichgewicht. Politische Ideen und staatsmännisches Handeln. Leipzig 1942; aufschlußreich auch *ders.*, Die Tradition der Großen Mächte. Stuttgart 1950, S. 79ff.

1787 deutlich geworden, als sich – im Zusammenhang mit den Plänen Kaiser Josephs II., die (seit 1713) Österreichischen Niederlande gegen Bayern einzutauschen und die Wittelsbacher Kurfürsten von München nach Brüssel zu verpflanzen – französische Ansprüche auf Holland geregt hatten. Pitt hatte solche Begehrlichkeiten damals durch ein 1788 vertraglich vereinbartes Zusammengehen mit Preußen energisch und erfolgreich abzuwehren gewußt.[43] Und noch über ein Jahrhundert später, Anfang August 1914, sollte sich die Haltung Englands von sehr ähnlichen Erwägungen leiten lassen: Bekanntlich traf die Londoner Regierung unter Premierminister Lloyd George den Entschluß zum Kriegseintritt erst in dem Augenblick, als deutsche Truppen durch die Verletzung der Neutralität Belgiens das vitale britische Sicherheitsinteresse an der Kanalküste unmittelbar berührten.[44]

Pitt konnte sich bei seinem nun mit großer Entschiedenheit eingeschlagenen und konsequent durchgehaltenen Kriegskurs auf die Billigung seiner Politik durch eine zeitweise geradezu erdrückende Mehrheit der Unterhausmitglieder verlassen, obgleich sein parlamentarischer Hauptgegenspieler Charles James Fox keine Gelegenheit versäumte, *Pitt's war* bei den Abgeordneten zu diskreditieren. Als leidenschaftlicher Anhänger der Französischen Revolution sah Fox in der gesamten britischen Kriegspolitik nichts weiter als einen großangelegten Versuch zur Unterdrückung der sich in Frankreich Bahn brechenden Freiheitsgesinnung und zur Erzeugung eines innenpolitischen Klimas, in dessen überhitzter Atmosphäre das Ministerium dann bequem zum Abbau mühsam errungener Freiheitsrechte und zur Wiedererrichtung eines vermeintlichen monarchischen Despotismus schreiten konnte.[45] Demgegenüber versuchten Pitts Anhänger, den Kriegskurs der Regierung durch umfangreiche publizistische Aktivitäten argumentativ zu rechtfertigen. Man hat errechnet,[46] daß allein in den 1790er Jahren, zwischen 1793 und 1802, über 400 separate Veröffentlichungen – Bücher, Abhandlungen und Pamphlete – Englands Teilnahme am Krieg verteidigten. Dabei flossen „ideologische“ und „realpolitische“ Interpretationen, wie sie von Burke

43 Den Zusammenhang zwischen bayerischem Tauschprojekt und holländischer Krise erörtert – neben der älteren Detailskizze von *John Holland Rose*, Great Britain and the Dutch question in 1787–1788. In: American Historical Review 14 (1908/09), S. 262–283 – jetzt vorzüglich *Jeremy Black*, British Foreign Policy in an Age of Revolutions, 1783–1793. Cambridge 1994, bes. S. 82 ff., 130 ff.

44 Darüber im größerern Rahmen *Horst Lademacher*, Die belgische Neutralität als Problem der europäischen Politik 1830–1914. Bonn 1971, S. 457–469.

45 Dieser „whiggistisch-antidespotischen“ Fehlinterpretation der Pitt'schen Politik gelten die Analysen von *John R. Dinwiddy*, Charles James Fox and the people. In: History, N. S. 55 (1970), S. 342–359, und *Richard E. Willis*, "A handful of violent people". The Nature of the Foxite Opposition, 1794–1801. In: Albion 8 (1976), S. 236–254.

46 So *M. Wagner*, England und die französische Gegenrevolution (wie Anm. 37), S. 261.

und von Pitt jeweils vorgegeben worden waren, zusehends ineinander. Der „burkeanisch" dominierten Sichtweise, die den unvereinbaren Gegensatz zwischen dem politischen System Alt-Englands und der sich seit 1789 etablierenden neuen Ordnung in Frankreich betonte und den Waffengang weiterhin als Kampf gegen die weltanschauliche Herausforderung duch die Revolution empfand, trat immer stärker eine macht- und interessenpolitisch orientierte Komponente zur Seite, die im Krieg gegen Frankreich willkommene Chancen für ein lohnendes Geschäft witterte. Ihren Verfechtern wogen die ökonomischen Marktvorteile, die dem britischen Handel infolge der kriegsbedingten Schwächung des französischen Konkurrenten winkten, jede militärische Anstrengung allemal auf.

Breite Unterstützung gewährte die britische Öffentlichkeit auch jenen vielbeschriebenen innenpolitischen Maßnahmen, die Pitt zur Flankierung seines außenpolitischen Kurses nach 1793 meinte treffen zu müssen, und die ein „whiggistisch" argumentierender Historiker wie Erich Eyck noch 1946 als ungerechtfertigte „Unterdrückungspolitik" brandmarkte.[47] Tatsächlich waren die in diesem Zusammenhang erlassenen Bestimmungen alles andere als leuchtende Signale britischer Freiheitsgesinnung. Die zweimalige Suspendierung der *Habeas-Corpus-Akte* (1794/95 und 1798 bis 1801) diente der Regierung Pitt ebenso zur Abwehr revolutionärer Umtriebe wie das seit 1799 geltende Verbot gewerkschaftlicher Zusammenschlüsse oder die ab 1792 (und bis 1799) mehrfach gesetzlich verschärfte Einschränkung der Presse-, Vereins- und Versammlungsrechte. Diese Repressionspolitik kann als Ausdruck eines von Furcht und Unsicherheit gleichermaßen diktierten Handelns gedeutet werden, als eine Art „nationaler Selbstschutz"[48], dessen Ziel die Bewahrung, nicht jedoch die Unterminierung der Verfassungsordnung des Inselreiches in ihrem überlieferten Bestand gewesen ist. Denn diese Verfassungsordnung sah sich seit Ende 1791 tatsächlich dem massiven Beschuß jenes *popular radicalism* ausgesetzt, der seine Anhängerschaft aus den halbgebildeten proletarischen Unterschichten der mittel- und nordenglischen Industrieregionen rekrutierte, und dessen teilweise sehr straff geführte Organisatio-

47 *E. Eyck*, Die Pitts und die Fox' (wie Anm. 2), S. 400; dazu allgemein *Clive Emsley*, British Society and the French Wars, 1789–1815. London 1979; zu den umstrittenen Methoden der Regierung vgl. ferner *ders.*, The Home Office and its Sources of Information and Investigation, 1791–1801. In: English Historical Review 94 (1979), S. 532–561.

48 So *Kurt Kluxen*, Geschichte Englands. Von den Anfängen bis zur Gegenwart. 4. Aufl. Stuttgart 1991, S. 494.

nen – die *Friends of Liberty*[49], die *Friends of Peace*[50] sowie andere „demokratische Clubs“[51] – eine grundlegende Umgestaltung der politischen und sozialen Ordnung Großbritanniens nach dem Vorbild des revolutionären Frankreichs anstrebten.[52] Der Agitation dieser *english Jacobins*[53] galt das Kriegsgeschehen lediglich als ein großangelegtes Manöver zur Stabilisierung des traditionellen Herrschaftsanspruchs der britischen Führungseliten.

Dieser relativ kleinen, jedoch ebenso engagierten wie lautstarken Gruppierung britischer Revolutionsbefürworter[54] stand nun allerdings die Majorität jener gegenüber, die als *popular loyalists* Pitts kriegspolitischen Kurs vorbehaltlos unterstützten. Organisiert in patriotischen Clubs, in royalistischen Assoziationen und vaterländischen Vereinen, die seit November 1792 in allen Teilen des Landes wie Pilze aus dem Boden schossen,[55] agierten

---

49 Vgl. maßgeblich *Albert Goodwin*, The Friends of Liberty. The English Democratic Movement in the Age of the French Revolution. London 1979, bes. S. 136–287, 332 ff., 387–415, 451–465; zur ideengeschichtlichen Verortung *Iain Hampsher-Monk*, Civic Humanism and Parliamentary Reform. The Case of the Society of the Friends of the People. In: Journal of British Studies 50 (1975), S. 70–89.

50 Vgl. *John E. Cookson*, The Friends of Peace. Anti-War Liberalism in England 1793–1815. Cambridge 1982; allgemeiner *Edward Royle/James A. Walvin*, English Radicals and Reformers 1760–1848. Brighton 1982, bes. S. 48–79.

51 Vgl. allgemein bereits *Simon Macoby*, English Radicalism 1786–1832. From Paine to Cobbett. London 1955; zum neueren Forschungsstand *Harry T. Dickinson*, British Radicalism and the French Revolution 1789–1815. London 1985, S. 16 ff.

52 Zur Ideologie der *radicals* vorzüglich *Günther Lottes*, Politische Aufklärung und plebejisches Publikum. Zur Theorie und Praxis des englischen Radikalismus im späten 18. Jahrhundert. München/Wien 1979, bes. S. 114–162, 275 ff.; *John R. Dinwiddy*, Conceptions of Revolution in the English radicalism of the 1790s. In: Eckhart Hellmuth (Hrsg.), The Transformation of Political Culture. England and Germany in the Late Eighteenth Century. Oxford 1990, S. 535–560; allgemeiner *Malcolm I. Thomis/ Peter Holt*, Threats of Revolution in Britain, 1789–1848. London 1977.

53 *James A. Walvin*, The English Jacobins 1789–1799. In: Historical Reflections 6 (1977), S. 91–109.

54 Edmund Burke zufolge vertraten die englischen *Jacobins* allenfalls ein Fünftel der „politischen Nation“ des gesamten Landes. Doch dieses Fünftel arbeite mit großer Energie am Totalumsturz der bestehenden Gesellschaftsordnung, während die zahlenmäßig weit überwiegende gegenrevolutionäre Mehrheit des Volkes in Indolenz und Apathie verharre; vgl. *Edmund Burke*, Letters on a Regicide Peace. In: Ders., The Works and Correspondences (wie Anm. 30), Bd. VIII. London 1852, S. 67 f.

55 Zur Organisationsstruktur dieser „loyalistischen“ Assoziationen und zu ihrer Einstellung gegenüber dem Ministerium Pitt vgl. im großen Rahmen *Eugene Ch. Black*, The Association. British Extra-parliamentary Political Organization, 1769–1793. Cambridge (Mass.) 1963; *Robert Hole*, British counter-revolutionary propaganda in the 1790s. In: Colin Jones (Hrsg.), Britain and Revolutionary France. Conflict, Subversion and Propaganda. Exeter 1983, S. 53–69; *Ian R. Christie*, Stress and Stability in late Eighteenth-Century Britain. Reflections on the British Avoidance of Revolution. Oxford 1984, bes. S. 156–182; *Emily Lorraine de Montluzin*, The Anti-Jacobins 1798–1800. The Early Contributors to the “Anti-Jacobin Review”. London 1988,

diese „Loyalisten" als gleichfalls sehr vernehmliche Fürsprecher einer antirevolutionären Massenbewegung. Ihre Wortführer waren fest entschlossen, die *old constitution* – und das hieß: die bestehenden Verhältnisse in Politik, Staat und Gesellschaft – kompromißlos zu verteidigen.[56] Das Einzugsfeld dieser *loyalists* umfaßte keineswegs bloß die sozialen Eliten des Landes. Es reichte bis weit in die Kreise des „einfachen Volkes", dessen konservative Grundstimmung sich in wachsendem Ausmaß zu einer allgemeinen Neuerungsfeindlichkeit verdichtete. „Reform" galt zusehends als ein Unwort, seine Verwendung wurde immer stärker mit der verhaßten politischen Entwicklung in Frankreich identifiziert.

Der von den Briten mehrheitlich gebilligte Krieg wurde zunächst, ab Februar 1793, im Bündnis mit Österreich und Preußen geführt, und Pitt hat sich in den folgenden Jahren intensiv darum bemüht, dieser Allianz einige Dauer zu verleihen. Doch schon die nur wenige Jahrzehnte zurückliegenden, in England besonders populären Erfolge Friedrichs des Großen über die Vielzahl seiner Gegner hatten die immensen militärstrategischen Strukturprobleme offenbart, denen jede Koalitionskriegführung im vorrevolutionären Europa des *Ancien Régime* unweigerlich ausgesetzt war. Dazu zählten nicht nur das Fehlen einer gemeinsamen politischen Zielsetzung und ein damit zusammenhängendes Überwiegen territorialer Sonderinteressen der miteinander verbündeten Partner, sondern auch deren immer wieder zutage tretender Mangel an taktischer Koordination und strategischer Kooperation.[57] Probleme dieser Art stellten sich auch den Alliierten von 1793, und

bes. S. 1–40; auf einzelne Zeitabschnitte beschränken sich die hinsichtlich ihrer Einschätzungen kontroversen Untersuchungen von *Austin V. Mitchell*, The Association Movement of 1792–3. In: Historical Journal 4 (1961), S. 56–77, und *Donald E. Ginter*, The Loyalist Association Movement of 1792–93 and British Public Opinion. In: Ebd. 9 (1966), S. 179–190.

56 Zur „loyalistischen" Ideologie vgl. die aufschlußreiche Studie von *Harry T. Dickinson*, Liberty and Property. Political Ideology in Eighteenth-Century Britain. London 1977, bes. S. 270–318, die nicht nur Parlamentsdebatten und Traktate führender Denker und Politiker auswertet, sondern auch die gesamte zeitgenössische politische Publizistik und Pamphletistik einschließlich zahlreicher Zeitungen, Zeitschriften, Flugblätter und Predigten berücksichtigt; vgl. auch *ders.*, Popular Conservatism and Militant Loyalism 1789–1815. In: Ders. (Hrsg.), Britain and the French Revolution, 1789–1815. London 1989, S. 113–121; *ders.*, Popular Loyalism in Britain in the 1790s. In: E. Hellmuth (Hrsg.) The Transformation of Political Culture (wie Anm. 52), S. 503–533; ferner *Robert R. Dozier*, For King, Constitution and Country. The English Loyalists and the French Revolution. Lexington 1983, bes. S. 26 ff., 55–75, sowie *Thomas Ph. Schofield*, Conservative Political Thought in Britain in Response to the French Revolution. In: Historical Journal 29 (1986), S. 601–622.

57 Vgl. *Walther Hubatsch*, Koalitionskriegführung in neuester Zeit historisch-politisch betrachtet. In: Schicksalfragen der Gegenwart. Handbuch politisch-historischer Bildung. Bd. 1. Tübingen 1957, S. 245–270; spezieller *Andreas Hillgruber*, Formverän-

Pitt hatte alle Hände voll zu tun, seine nicht eben verläßlichen festländischen Bundesgenossen zusammenzuhalten. Gelungen ist ihm dies nicht. Preußen schied 1795 (mit dem Frieden von Basel), Österreich 1797 (mit dem Frieden von Campo Formio) aus der Koalition aus – beide Mächte nahmen dafür faktisch die Preisgabe des linken Rheinufers an Frankreich in Kauf, was zu verhindern für Pitt ja gerade der Grund gewesen war, in den Krieg einzutreten. Ein britisch-französischer Friedensschluß kam unter diesen Umständen auf keinen Fall in Betracht. Und so standen sich denn eine zeitlang England und Frankreich allein gegenüber, bevor der Seesieg der Briten bei Abukir im August 1798 die Voraussetzung für den Abschluß der Zweiten Koalition gegen Frankreich schuf. Nun war England mit Österreich und Rußland verbunden, und nun besaß dieses Bündnis auch eine gemeinsam formulierte politische Zielsetzung. Sie umfaßte die Zurückführung Frankreichs auf die Grenzen von 1792, die Beseitigung der revolutionären Regierung in Paris sowie die Verpflichtung auf Wiedereinsetzung des bourbonischen Königshauses.[58]

Britischen Traditionen gemäß hat Pitt „seinen" Krieg, wie dies bereits sein Vater, William Pitt der Ältere, in den Jahren von 1757 bis 1761 getan hatte,[59] vornehmlich als einen überseeischen Handels-, See- und Kolonialkrieg mit durchgängig antifranzösischer Tendenz geführt. Und er hat hier, wiederum in Anknüpfung an das Werk des Vaters, seine großen, dauerhaften Erfolge errungen, die für Großbritanniens Weltstellung bis zur Mitte des 20. Jahrhunderts Bestand haben sollten.[60] Sein Engagement im kontinentaleuropäischen Landkrieg beschränkte sich demgegenüber darauf, die Verbündeten mittels diplomatischer Initiativen und großzügig bemessener Subsidienzahlungen ideell und finanziell zu unterstützen. Auch dies erwies sich als ein probates, bereits vom Vater mit einigem Erfolg praktiziertes Verfahren zur Aufrechterhaltung alliierten Konsenswillens. In *zweierlei* Hinsicht besaßen des jüngeren Pitts Verdienste um das *Empire*, um die Neufundierung, Stär-

derung in der Koalitionskriegführung in der Epoche 1792/1815 – Ein Aufriß (1966). Wiederabgedruckt in: Ders., Deutsche Großmacht- und Weltpolitik im 19. und 20. Jahrhundert. Düsseldorf 1977, S. 12–34.

58 Die gleichwohl erheblichen Gegensätze zwischen den Allianzpartnern und ihre letztlich unüberbrückbaren politischen Differenzen betont und analysiert scharfsichtig *Paul W. Schroeder*, The Collapse of the Second Coalition. In: Journal of Modern History 59 (1987), S. 244–290.

59 Darüber zusammenfassend *Gottfried Niedhart*, Handel und Krieg in der britischen Weltpolitik 1738–1763. München 1979, bes. S. 62 ff., 100 ff., 121–138, 160 ff.

60 Für den Zusammenhang sehr instruktiv *Peter J. Marshall*, The Eighteenth-Century Empire. In: Jeremy Black (Hrsg.), British Politics and Society from Walpole to Pitt 1742–1789. London 1990, S. 177–200; grundlegend noch immer *Vincent T. Harlow*, The Founding of the Second British Empire, 1763–1793. Bde. 1–2. London 1952–1964.

kung und Erweiterung des britischen Kolonial- und Weltreiches, eine besondere Relevanz.

Erstens: Zu den langfristig nachwirkenden Erfolgen Pitts im See- und Kolonialkrieg gegen Frankreich zählte – neben der Eroberung Trinidads, Maltas, des Kaplandes, Ceylons und der Sunda-Inseln – vor allem der triumphale Sieg der britischen Flotte über die französisch-spanische Armada bei Kap Trafalgar im Oktober 1805 während des Dritten Koalitionskrieges, drei Monate vor Pitts Tod. Dieser Seesieg machte die befürchtete Invasion Napoleons I. in England definitiv unmöglich. Der Alptraum einer solchen Invasion, die fixe Idee einer Landung französischer Truppen an der Westküste Englands, hatte Pitt seit dem Abschluß des nicht mehr von ihm verantworteten und daher mit umso größerer Skepsis betrachteten Friedens von Amiens (27. März 1802) aufs Stärkste beschäftigt. Bei Parlamentsdebatten warb er mit unermüdlichem Eifer für eine Intensivierung des Küstenschutzes, verlangte beschleunigte Maßnahmen zur Landesverteidigung und riet zur Anlage von Feldschanzen, um Londons Sicherheit zu gewährleisten. Er selbst hatte damit begonnen, von seinem Wohnsitz *Walmer Castle*, aus ein örtliches Freiwilligenaufgebot auszubilden, und er beabsichtigte ernsthaft, im Mobilmachungsfall an die Spitze seiner Leute zu treten.[61]

Alle im Krieg errungenen Erfolge Pitts waren indes nur eine Art Fortsetzung der vom ihm bereits in Friedenszeiten eingeleiteten Politik eines „imperialen Neubeginns". Diese Politik begann unmittelbar nach seinem Amtsantritt mit der kostspieligen Wiederinstandsetzung der britischen Marine, deren Qualität durch den Krieg in Nordamerika stark gelitten hatte. Pitt setzte damit ein weithin sichtbares Signal, das als zukunftweisendes Symbol für die angestrebte Konsolidierung der britischen Weltmachtstellung verstanden werden kann. *Drei* Geschehensregionen standen dabei im Mittelpunkt seiner diesbezüglichen Aktivitäten. – *Australien*, formell erst 1770 für die britische Krone in Besitz genommen, galt ihm ab 1786 als bevorzugtes Ziel siedlungspolitischer Unternehmungen, und es ist Pitt gewesen, der den fünften Kontinent als Sträflingskolonie zur Verschickung heimischer Häftlinge erstmals in großem Umfang zu nutzen gewußt hat. – *Kanada* verdankte seiner Regierung eine umfassende Verwaltungsneugliederung. Durch ein vom britischen Parlament 1791 verabschiedetes Verfassungsgesetz wurde die Provinz Quebec, deren Bevölkerung durch die Flucht tausender aus den USA vertriebener englandtreuer *Empire loyalists* stark angewachsen war, in das französischsprachige *Lower Canada* und das englischsprachige *Upper Canada* geteilt. Jeder der beiden Regionen wurde eine eigene Volksvertretung und begrenzte Verwaltungsautonomie zugestanden. Damit waren

61 Vgl. *J. H. Rose*, Der jüngere Pitt (wie Anm. 2), S. 218 f.

deutliche Lehren aus dem nur wenige Jahre zurückliegenden nordamerikanischen Desaster gezogen. – *Britisch-Indien* schließlich war bereits 1784 zum Gegenstand einer von Pitt im Parlament eingebrachten Gesetzesvorlage geworden.[62] Der *East India Act* übertrug die Oberaufsicht der ins Gerede gekommenen *East India Company*[63] auf das Parlament bzw. das Kabinett. Die *Company* behielt das Recht, vor Ort selbständige Entscheidungen zu treffen, doch alle ihre Handlungen waren hinfort der Kontrolle der Regierung in London unterstellt.[64] Der damit von Pitt eingeleitete Neuanfang eröffnete eine Ära britisch-indischer Staatspolitik, die in ihren Grundzügen bis zur Übernahme aller von der *Company* verwalteten Territorien durch die britische Krone 1858 Bestand haben sollte.

Zweitens: Eine bruchlose Erfolgsgeschichte ist des jüngeren Pitt Engagement im Dienst der britischen Reichsidee allerdings nicht gewesen. Als wunder Punkt erwies sich vor allem seine Politik gegenüber Irland.[65] Dessen staatsrechtlich unklare Stellung zu Großbritannien wünschte er durch die Vereinigung der beiden Parlamente von Westminster und von Dublin, also durch eine *Union* nach dem Vorbild der Vereinigung Englands mit Schottland 1707,[66] endgültig zu definieren. Pitt dachte dabei an echte Gleichbe-

62 Dazu die älteren Spezialstudien von *Cyril H. Philips*, The East India Company "Interest" and the English Government, 1783–4. In: Transactions of the Royal Historical Society, 4th series, 20 (1937), S. 83–101; *ders.*, The New East India Board and the Court of Directors. In: English Historical Review 55 (1940), S. 438–446.

63 Auf Betreiben Edmund Burkes hatte sich das *House of Commons* 1787 zur Eröffnung eines von starkem Publikumsinteresse begleiteten Gerichtsverfahrens gegen den Gouverneur von Bengalen und ersten *governor-general* von Ostindien, Warren Hastings, entschlossen. Ihm wurde vorgeworfen, die rücksichtslos ausbeuterische Kolonialpraxis der *East India Company* befördert und gedeckt zu haben. Obwohl der Prozeß nach siebenjähriger Dauer 1795 mit einem Freispruch für den Angeklagten endete, waren im Verlauf des Verfahrens doch krasse Mißstände der britischen Kolonialherrschaft in Indien zutage getreten, die nach Abhilfe verlangten; dazu ausführlich und mit detaillierten Schilderungen der Rolle Pitts im Verhandlungsgeschehen *Peter J. Marshall*, The Impeachment of Warren Hastings. London 1965, bes. S. 39–65.

64 Die beste Studie über die innere Entwicklung der *Company* und ihr Verhältnis zur Regierung stammt noch immer von *Cyril H. Philips*, The East India Company, 1784–1834. 2. Aufl. Manchester 1961; wichtig weiterhin *Lucy S. Sutherland*, The East India Company in Eighteenth Century Politics. Oxford 1952; allgemeiner und gut dokumentierend *Peter J. Marshall*, Problems of Empire. Britain and India, 1757–1813. London 1968; zur Forschungsdiskussion *ders.*, British Expansion in India in the Eighteenth Century. A historical Revision. In: History, N. S. 60 (1975), S. 28–43.

65 Zur krisenhaften Zuspitzung des britisch-irischen Verhältnisses im späteren 18. Jahrhundert vgl. die Untersuchungen von *Edith M. Johnston*, Great Britain and Ireland, 1760–1800. A Study in Political Administration. Edinburgh 1963, und *Robert B. McDowell*, Ireland in the Age of Imperialism and Revolution, 1760–1801. Oxford 1979.

66 Vgl. *Patrick W. J. Riley*, The Union of England and Scotland. A Study in Anglo-Scottish Politics of the Eighteenth Century. Manchester 1978.

rechtigung, und das hieß konkret: an die Aufhebung der bisher herrschenden politischen Rechtlosigkeit der irischen Katholiken mittels Gewährung des Wahlrechts. Dieses emanzipatorische Vorhaben stieß jedoch auf den erbitterten Widerstand König Georgs III., sodaß der *Irish Act of Union* – nach massiver Bestechung der irischen Oligarchen – zwar im August des Jahres 1800 in den beiden Parlamenten von Westminster und von Dublin die jeweils erforderlichen Mehrheiten erhielt.[67] Doch Pitts eigentliches Ziel, durch eine Politik der Versöhnung gegenüber den Katholiken die irische Unterstützung für den nicht endenwollenden Krieg gegen Frankreich zu erlangen, ist nicht erreicht worden. Die Katholikenemanzipation im Vereinigten Königreich erfolgte bekanntlich erst 1829, dreiundzwanzig Jahre nach Pitts allzu frühem Tod.[68] Es zeugt von Standhaftigkeit und Überzeugungstreue, daß der Premierminister die Nichtgewährung der von ihm ausdrücklich versprochenen irischen Katholikenemanzipation durch König Georg III. zum Anlaß nahm, um am 3. Februar 1801, nach siebzehnjähriger Amtszeit, zu demissionieren.[69] Seine dann noch einmal von Mai 1804 bis Januar 1806 währende zweite Amtszeit stand vollkommen im Schatten des Kampfes gegen das Hegemonialstreben des nun kaiserlich gewordenen Frankreichs Napoleons I.

IV.

William Pitt der Jüngere hat in den beiden Jahrzehnten seiner aktiven politischen Wirksamkeit, die beinahe die Hälfte seines kurzen Lebens einnahmen, ein staatsmännisches Werk von erstaunlicher Konsistenz geschaffen. Und er hat diese Werkschöpfung unter erschwerten Bedingungen, zudem in konfliktreichen Kriegs- und Krisenzeitungen, vollbringen müssen. Während seine Zeitgenossen – Friedrich der Große und Joseph II. ebenso wie Napoleon Bonaparte – ihre Ziele in relativer Selbstherrlichkeit vorantreiben konnten und sie, zumindest was die beiden letzteren betrifft, keineswegs mit dauerhaftem Erfolg zu realisieren vermochten, lagen beim jüngeren Pitt die Dinge anders.

---

67 Dazu eingehend *Geoffrey C. Bolton*, The Passing of the Irish Act of Union. A Study in Parliamentary Politics. London 1966, S. 53 ff., 67 ff., 185–222; zur Vorgeschichte *James Kelly*, The Origins of the Act of Union. An Examination of Unionist Opinion in Britain and Ireland, 1650–1800. In: Irish Historical Studies 25 (1987), S. 236–263.

68 Darüber in größerem Rahmen *George I. T. Machin*, The Catholic Question in English Politics 1820–1830. Oxford 1964, bes. S. 157–178.

69 Für die Zusammenhänge aufschlußreich *Piers G. Mackesy*, War without Victory. The Downfall of Pitt 1799–1802. Oxford 1984; zu den Einzelheiten der Demission ferner die Quellenstudie von *Richard E. Willis*, William Pitt's Resignation in 1801. Reexamination and Document. In: Bulletin of the Institute of Historical Research 44 (1971), S. 239–256.

Er, der von vielen Seiten bedrängte Erste Minister eines sich zusehends parlamentarisierenden Staatswesens, angefochten von einer oftmals bedrohlich aggressiven Unterhausopposition, und in ständiger Spannung zu einem unberechenbaren und undankbaren Monarchen, hat gleichwohl das Kunststück vollbracht, das europäische Kräftegleichgewicht gegen die Hegemonialgelüste des französischen Aggressors – revolutionärer wie imperialer Provenienz gleichermaßen – erfolgreich zu verteidigen. Und er hat darüber hinaus, überlieferten Traditionen britischer Machtpolitik ebenso folgend wie solche Traditionen durch sein Werk erst selbst begründend, diese auf Kontinentaleuropa bezogene Gleichgewichtsdoktrin dazu genutzt, die überseeische Expansion und maritime Welt-Vorherrschaft seines eigenen Landes mit großem Geschick ins Werk zu setzen. Im Windschatten des kontinentalen Geschehens, weitgehend unberührt von allen innereuropäischen Umwälzungen und Verwerfungen, hat er *Englands Rise to Greatness* (Stephen Baxter) maßgeblich befördert. Solche staatsmännische Meisterschaft verdient den Respekt der Historiker – der englischen allemal, und wohl auch der deutschen.

Derek Beales

Joseph II: Wars Intended, Prevented and Unexpected

In a Viennese pamphlet of 1788, entitled *A Word in Confidence about the Turkish War*, *Ein Wort im Vertrauen über den Türkenkrieg*, it was claimed that in Britain, with its limited monarchy, the people would never indulge the ruler's dynastic ambitions and personal whims, and hence that a war such as the Austrians' war against the Turks, which began in February 1788, would for Britain be unthinkable[1]. This was to take both a rosy view of the position in Britain and a jaundiced view of the Turkish war. For no one concerned with policy-making in the Austrian Monarchy could possibly suppose that the public or the Estates of its numerous provinces would have even the slightest part in the state's decision to engage in war. It is true that remarkably strong criticisms of the Turkish war and its conduct were found in many pamphlets and newspapers published within the Monarchy, which in the central lands at least were rarely suppressed or censored. But, except for anodyne government-inspired bulletins and announcements, there was almost no information in the public domain about the *making* of foreign and military policy, especially before the decisions were taken[2]. In the historiography of the Monarchy's foreign policy there is certainly no need for a '*return* of great men' or of 'personalities'. They have never left it and never could leave it. The monarchs, successively Maria Theresa, Joseph II and Leopold II, were unquestionably absolute rulers.

Further, whereas in domestic matters there was an elaborate administrative structure which could influence and obstruct the ruler, headed by the Council of State (*Staatsrat*) and including many provincial governments, in foreign policy there was only one body directly concerned, the *Staatskanzlei*, the State Chancellery, and the only serious restrictions on his freedom action were the state of the army and the finances. In virtually all his

1 This pamphlet is usefully discussed in *E. Wangermann*, Die Waffen der Publizität, Vienna 2004, 176–9.

2 See *Derek Beales*, Joseph II, vol. II: Against the World, 1780–1790, Cambridge 2009, esp. 589–82, 625, and for Belgium, 616–19.

provinces the ruler had to request the Estates to grant him taxes. This requirement could in rare cases be crucial in the conduct of wars, but it was doubtful whether the Estates were constitutionally entitled actually to refuse the ruler's demands, even in Belgium and Hungary, and, so far as I know, they never ventured to discuss foreign policy. Count Karl von Zinzendorf, who from 1782 was in charge of balancing the Monarchy's budget, was instructed by Joseph that there needed always to be enough money available for one year's campaign, but he was not consulted about foreign policy and military needs in advance. Only when the decisions on these matters had been taken was he told that funds had to be found to implement them[3]. The only body concerned with foreign policy which was comparable in standing with the *Staatsrat* in domestic affairs was the Privy Conference or *Conferenz*. But, though it had not been abolished, this committee was never called together or (it seems) consulted between the Seven Years War and the beginning of 1790, when Joseph, acknowledging that he was no longer capable of governing, revived it[4].

In Joseph's sole reign from 1780 to 1790, as in the last years of Maria Theresa's reign, the only persons who had a major influence on foreign policy were the following: pre-eminently the state chancellor, prince Kaunitz, who had held this office since 1753; a few high foreign office officials and one or two ambassadors such as count Mercy in Paris and, later, count Ludwig Cobenzl in St Petersburg; and, increasingly, his cousin, count Philipp Cobenzl, made vice-chancellor, that is, deputy to Kaunitz, as a *protégé* of Joseph II in 1779. In addition field-marshal count Lacy, though no longer the official head of the War Department, was treated as such by Joseph, who regularly sought and often took his advice on issues of foreign as well as military policy[5]. It is interesting also that count Rosenberg, the chief court chamberlain and the man put in charge of the opera, was also the person Joseph had in mind as Kaunitz's eventual successor. It looks as though much more than the opera was discussed between Rosenberg and the emperor during their daily meetings early each morning, at which the count,

---

3 See *P. G. M. Dickson*'s writings, esp. his Count Karl von Zinzendorf's "New Accountancy", the Structure of Austrian Government Finance in Peace and War, 1781–1791, International History Review 29 (2007), 22–56; *Beales* (note 2), esp. 374, 381.

4 Ibid., 630–2. On the earlier history of the Conferenz, *P. G. M. Dickson*, Finance and government under Maria Theresia, 1740–1780, 2 vols., Oxford 1987, vol. I, 340–1; vol. II, 449, 452; *E. Matsch*, Geschichte des auswärtigen Diensts von Österreich (-Ungarn), 1720–1920, Vienna 1980, 28–31; *M. Hochedlinger*, Krise und Wiederherstellung, Berlin 2000, 309–14.

5 On Lacy *Beales* (note 2), 37; on Philipp Cobenzl's appointment *Derek Beales*, Joseph II, vol. II: In the shadow of Maria Theresa, 1741–1780, Cambridge 1987, 426.

in Da Ponte's words, handed Joseph his shirt[6]. It must be significant that Lacy and Rosenberg were the only men who formed part of Joseph's regular evening conversations under the aegis of the five great ladies known as the *Dames*. Kaunitz, Lacy and Rosenberg were gazetted throughout the period as members of the *Conferenz*. The only other member was prince Starhemberg, who was much less influential, having been sent to govern Belgium in 1770 and stayed there until 1783[7]. I shall discuss later the role of Joseph's brother, acknowledged heir and successor, grand-duke Leopold of Tuscany.

The difference between the making of internal and external policy was highlighted by the fact that Philipp Cobenzl and Lacy had immediate access to Joseph by a separate staircase from that which was available to domestic officials[8]. Kaunitz of course had the same right in principle but under Joseph seems never to have exercised it. Notoriously, the emperor had got into the habit of going to the chancellor's office to see him rather than *vice versa*. This was partly a question of Kaunitz's age – he had been slowing down and had been described as "*still having excellent judgment, but not as good memory*"[9]. Even more extraordinarily, in the last two years of the reign Joseph and Kaunitz never met at all, because the chancellor refused to see anyone who was ill and the emperor acquiesced in this ban when he returned from campaign at the end of 1788 stricken with tuberculosis[10]. Thereafter Philipp Cobenzl or, failing him, Baron Spielmann, had to act as their go-between.

It does not necessarily follow from the fact that so few individuals were able to influence foreign policy significantly that the policy adopted was the product, as the author of the 1788 pamphlet claimed, of the ruler's dynastic ambitions and personal whims. Kaunitz at least had developed an impressive theoretical framework for identifying the interests of the state rather than those of the dynasty, let alone the whims of the ruler. His thinking greatly influenced the attitudes of Maria Theresa, Joseph II and Leopold II[11]. It seems to me that Joseph's views, though rather different from Kaunitz's and less coherent, were also usually founded on rational consideration of the Monarchy's interests.

---

6 *Da Ponte*, Memoirs, hg. von L. A. Sheppard, London 1929, 132. *Beales* (note 2), 37–8. Zinzendorf's diary is full of information derived from Rosenberg about J.'s foreign and other policies.
7 On Starhemberg *Beales* (note 2), 343–5.
8 Sec ibid., 434–5.
9 Philipp Cobenzl to Ludwig Cobenzl, 11 Oct. 1787 (*A. Beer und J. von Fiedler*, Joseph II. und Graf Ludwig Cobenzl, 2 vols., Vienna 1901, Vol. II, 20).
10 *Beales* (note 2), 627.
11 See *H. Klueting*, Die Lehre von der Macht der Staaten, Berlin 1986.

It is well known that this period was unusual in that the wars of western Europe and eastern Europe seemed to be kept distinct from each other[12]. But, since the Monarchy included both Belgium and many eastern provinces, its rulers had always to consider the relationship between East and West. And, although the East is usually regarded as the area of major interest during Joseph's reign, and it certainly generated the largest-scale war, in fact three of the five actual or possible wars that I am going to discuss related to the Low Countries, and the position of the Belgian provinces was always of major concern to both Joseph and Kaunitz.

In the Peace of Utrecht at the end of the War of the Spanish Succession the Powers had very cunningly – perhaps more cunningly than they realised – saddled Vienna with the former Spanish duchies that today more or less compose Belgium, provinces separated by hundreds of miles from the Monarchy's core lands. In the same Treaty the Powers had further stipulated that the river Scheldt was to be closed to the commerce of the Belgian provinces. Joseph's first initiative in foreign affairs during his sole reign concerned this question[13]. Before the end of 1780, when he had only just succeeded his mother, the British government, now at war with the Dutch, put out a feeler to Austria offering to abandon its long-standing opposition to the opening of the Scheldt. This feeler was intended to pave the way for a proposal to revive the Anglo-Austrian alliance which had existed for the best part of a century before the Diplomatic Revolution of 1756. Joseph wrote to Kaunitz that he was tempted by the British proposal. The emperor resented the restriction to his sovereignty that the closure of the Scheldt represented, and he also hankered after the 'old' English alliance.

Kaunitz's whole career had been built on the switch of Austria's alliance from Britain to France in 1756. When Joseph showed sympathy with the British initiative over the Scheldt, Kaunitz sent him a curt and devastating reply in which he claimed that to make this deal with Britain would rupture the alliance with France,

> "*which alone, so long as it lasts, enables us to stand up to the King of Prussia and even to the Ottoman Empire if one or the other decides to attack us – something absolutely impossible in any other case than in the present state of affairs.*"

Not only would it destroy Austria's 'system of alliances', but also it would "*unquestionably bring upon us a general war*". This latter prophecy at least

---

12 See, e.g. *H. M. Scott*, The Birth of a Great Power System, 1740–1815, Harlow 2006, esp. chs. 5–8.

13 For this and the next para. See *Beales* (note 2), esp. 111–13.

seems unconvincing: others, such as Keith, the British envoy to Vienna, thought that France could not have risked war against Austria while engaged in fighting Britain over America. But Joseph gave in, this time round. So in this case it was Kaunitz who claimed to have prevented a war that Joseph's policy might have precipitated.

For the next few months the two men were preoccupied with trying to conclude an alliance with Russia, and therefore with the affairs of the East. Using one of his trademark weapons, travel, Joseph had visited Russia and softened up Catherine II in 1780, during his mother's lifetime[14]. Later it was Joseph who insisted, against Kaunitz's initial objections, on sidestepping problems over the precedence of the emperor and empress and making the treaty by means of almost identical handwritten letters. It was agreed that the existence of the alliance would be denied, concealed from all but a handful of ministers and kept out of the ordinary archives. This was Cabinet diplomacy *par excellence* – a secret, personal deal between a great men and a great woman.

In the alliance made with Catherine II of Russia in May-June 1781 Joseph agreed to back her with forces equal to those she herself deployed in two possible situations, first if the Turks reneged on the promises concerning their Greek Orthodox subjects which they had made in the peace of Kutchuk-Kainardji that had ended the last Russo-Turkish War in 1774, and secondly if the Turks attacked Russia. In her turn Catherine promised that, if during a war with Turkey in which Austria was supporting Russia Austria was attacked by another Power, Russia would support her against that country "*with all its forces*".

One of the commonplaces of the historiography of Joseph's reign is that his brother Leopold was not kept informed about the alliance treaty and, when eventually he found out about it, strongly disapproved of it. In Arneth's published collection of the letters between the two brothers, there is no letter telling Leopold that the alliance had been sealed[15]. Madariaga and Wandruszka accept that none existed. But I found in the archive just such a letter to his brother dated 20 May, enjoining the strictest secrecy, and telling him, slightly inaccurately, the terms of the alliance he had concluded.

Leopold had previously written to Joseph in February, when he had been prematurely told that the emperor was hoping to negotiate a Russian alliance, in these terms:

---

14 See *Beales* (note 5), 431–8.

15 On the making of the alliance see *I. de Madariaga*, The Secret Austro-Russian Treaty of 1781, Slavonic and East European Review XXXVIII (1959), 114–145; *A. Wandruszka*, Leopold II., 2 vols, Vienna 1964–5, vol. II, 41–5; *Beales* (note 2), 113–23. On J.'s letter of 20 May 1781 to L. and L.'s of 28 Feb. see ibid. 121–2 and nn.

> *"I compliment you most sincerely on the wisdom and prudence with which during your journey to St Petersburg you prepared and led, and now have succeeded in completely winning over, the Empress of Russia, opening her eyes about Prussia, detaching her from it, and finally achieving the great coup of concluding an offensive and defensive alliance, the most natural, the most advantageous, useful and suitable in every way and in all the possible circumstances to our dynasty."*

Leopold's agreement that a relative of Catherine should be the bride for his eldest son, Francis, was crucial to the alliance. Leopold later repeated his approval of the alliance in his private and secret account of the exhaustive discussions about policy that he had with Joseph in Florence and Vienna in 1784. When the time was approaching for the alliance to be renewed, in 1789, the emperor insisted, over the objections of Kaunitz, that Leopold be consulted first and give his approval, which he did[16]. Leopold was very reluctant to disagree openly with his brother, especially because he knew that only Joseph could provide prestigious roles for his huge family. So it was Leopold rather than Joseph who put dynastic considerations first. But it is hard to doubt the sincerity of his support for this alliance. And what is absolutely certain is that Leopold was in fact kept astonishingly well informed. For example, he was told about the emperor's plan to exchange Bavaria for much of Belgium several months before Joseph broached the issue with Kaunitz and Lacy and so made it official policy[17]. He was given many opportunities to oppose Joseph's policies but hardly ever did so.

Joseph knew that Catherine had vast plans to dismember the Turkish lands, which she put on paper in a letter of September 1782. His response in November was to suggest a long series of territorial gains that he said would suit Austria, such as Belgrade and the Venetian *terrafirma*, Istria and Dalmatia. But it seems clear that he was making such inflated claims only because he regarded Catherine's as even more 'fantastic' or 'chimerical'. Joseph originally put into his letter listing these possible annexations the following statement:

> *"I regard it as the greatest principle of state to aim at rounding off rather than extending a Monarchy which is already big enough."*

Kaunitz had him remove this too revealing and too modest remark, but I think it represented Joseph's true opinion.

Catherine had dangled before Joseph even greater gains, especially in Italy, and she replied to him expressing disappointment that he was aban-

16 On the renewal in 1789 ibid. 590–1.
17 *Wandruszka* (note 16), 78.

doning "*a useful project, grand and worthy of Caesar*"[18]. But when it appeared that war in the East might be imminent, as Catherine complained of Turkish breaches of the last treaty with Russia and the Turks began to stir, Joseph mobilised his army – as a threat both to the Turks and to Prussia. Catherine was already treating the Crimea as a sort of protectorate. Now, early in 1783, she annexed it, without informing Joseph beforehand. By the time that news of this annexation reached Vienna, Joseph had already left the city *en route* to inspect Hungary, Transylvania and Galicia. Kaunitz pursued him with letters urging him in the strongest terms to seize the opportunity of going to war against the Turks in order to make comparable gains and win for himself 'glory' – Kaunitz actually used the English word. The emperor, writing from a series of remote locations, refused point-blank. He told Leopold:

> "*Prince Kaunitz absolutely wanted me to start a war at once by marching into Moldavia and Wallachia, or by* [denouncing existing treaties with Turkey] … [If I had done so,] *I should have the whole mass of the Turkish army on my back, and very probably the king of Prussia and perhaps even the Bourbon Courts in Belgium, on the Rhine and in Italy – and at the end of it a miserable piece of desert in Bosnia or Serbia for which I've risked everything. I cannot conceive how this intelligent man could get all this into his head, but I had to give a positive order and send back several prepared instructions to make him give in.*"

In this case Joseph unquestionably prevented a war that many expected to occur and that Kaunitz strongly urged him to start[19].

By this time, however, Joseph was stirring up trouble with the Dutch. He demanded that they should negotiate with him about his so-called grievances, namely, about the opening of the Scheldt and the cession of some of their territory, particularly the detached city of Maastricht. He saw the Dutch as an easy target and treated them rather as Hitler treated Czechoslovakia, provoking them by infringements of their frontiers and suggesting that if they did not satisfy him he would invade. This went on for over a year before Joseph declared in October 1784 that a state of war existed between the Monarchy and the Republic. The whole bizarre episode can only be explained by what he had earlier told Leopold, that its main purpose was

> "*to worry France and draw the attention of Europe to that area, so that the Courts would not pay attention to the planned acquisition of*

---

18 *Beales* (note 2), 376–82.
19 Ibid., 382–3.

> *Bavaria and would believe that ... Vienna had a strong desire to expand the* [Austrian] *Netherlands and their commerce, when its* [true] *intention was to give them up ... and it does not intend to insist on* [annexing] *Maastricht ... But it will absolutely demand the free navigation of the Scheldt, considering this achievement ... glorious."*

But the Dutch proved much less feeble than Joseph expected, his army was frustrated by their opening the dikes and flooding the frontier districts, and France gave them strong diplomatic support. In the end, in the treaty of Fontainebleau at the end of 1785, Joseph gave up almost all his claims in return for an indemnity of 9,500,000 francs, half paid by France. He achieved no territorial gain, the Scheldt was not opened to commerce, and France prevented the Bavarian exchange by paying a huge bribe to the next heir to the electorate, the duke of Zweibrücken, outtrumping the inducements offered by Joseph and Catherine. This preposterous Dutch War was the only war actually got up by Joseph II[20].

When the Turks declared war on Russia in August 1787, Joseph accepted that he was bound by his alliance with Catherine, as he had not been in the circumstances of 1783, to come to her support. He was reluctant, because he was on the point of sending troops to Belgium to suppress a rebellion there against his rule, and the need to fight the Turks prevented him sending enough. This war against the Turks was not a war intended by Joseph II, and it was an unexpected war in one sense: almost no one had dreamed in 1781 that a warlike party would come to power in Constantinople and that they would dare to attack Russia. But in another sense it had been, if not expected, allowed for in the Austro-Russian alliance treaty. However, Joseph delayed his declaration of war longer than the three months mentioned in the treaty, and he launched two strikes against the Turkish fortress of Belgrade before the two Powers were officially at war, both of them pathetic failures[21].

Whether he needed to commit almost his entire army, nearly 300,000 men, to the first campaign of the war is a much disputed question. But even if only 30,000 troops had been promised to assist Russia in this eventuality (which I don't think is true), it was inescapably necessary for him to deploy a much bigger force because his frontiers with the Ottoman Empire were so long.

As had happened in the War of the Bavarian Succession, once he joined his armies at the front he started whingeing, and writing letters to Vienna telling Kaunitz how desperate the situation was and how ill he was, requiring him, if at all possible, to engineer a peace – to which Kaunitz cruelly responded that his mother had never allowed herself to be so defeatist. But Joseph insisted

20 Ibid., 374–6, 390–8.
21 Ibid., 555–63.

and Kaunitz eventually obeyed[22]. These peace negotiations, however, bore fruit only under the aegis of Leopold II, who succeeded him in February 1790.

The last war I want to consider is that between Joseph and the army fielded by the Belgian rebels against his rule in the latter part of 1789. He certainly did not intend there to be a rising against his rule, but he unquestionably brought it on by his insistence on extremely unpopular ecclesiastical and administrative reforms leading to his quashing the constitutional charter of Brabant, the *Joyeuse Entrée,* in June 1789. With the connivance of the Dutch, the Prussians and the revolutionaries in the independent bishopric of Liège, a substantial force was gathered on the frontiers of Belgium, which then invaded it. Before the end of the year his troops had been driven out of all the Belgian provinces except Luxembourg, and before he died an independent revolutionary government was in control of the country. Kaunitz, whose immensely powerful position included responsibility for the domestic government of Belgium, had protested strongly against Joseph's measures, but had been overruled[23].

So, to summarise, Kaunitz had dissuaded Joseph early in 1781 from taking a step which he claimed would lead to war. In 1783 Joseph had rejected Kaunitz's strong advice to join his new ally Russia in a war against the Turks. But he was already planning a war against the Dutch which broke out in the following year. He joined his ally Russia in fighting the Turks from 1788 onwards, unwillingly but because the terms of his alliance required it. He had certainly not intended that Belgium should revolt against his rule and simply could not believe that his troops could be so easily driven out of the country late in 1789, just after the great triumph of his arms in the East, the capture of Belgrade. A great personality, Joseph, had provoked this war, but the rebel army had been put together and funded by relatively humble men and women, some nobles, some abbots, students, lesser clergy, lawyers, demonstrators, all stirred up by journalists. But once Leopold succeeded and abandoned his brother's reforms, the country was easily reconquered.

In conclusion, Joseph certainly hoped to gain territory, but preferably by partition, exchange or bullying without war. He was not nearly so inclined to deploy his army as has been assumed. He was sometimes less belligerent than Kaunitz. When he found himself committed by his Russian alliance to war, he was reluctant and slow to act. But that war he seemed to be winning late in 1789. The war that he could not win was the war he himself

22 Ibid., 568–74.

23 On J.'s policies in Belgium and the rebellions against them ibid., 499–507, 512–25, 604–05, 610–22.

had provoked by his despotic policies in Belgium. The other Powers were prepared to support Austrian rule in Belgium under the Treaty of Utrecht only if its rulers preserved the provinces' constitutions. Kaunitz lamented in December:

> *"It is frightful that despotic obstinacy should have put this fine Monarchy into its present condition. It is distressing to me not to be able to justify myself publicly, but I am to some degree consoled by the knowledge that the whole public is convinced that everything that has happened is entirely due to the limited attention that has been paid during this reign to my advice and to my courageous representations on all occasions"*[24].

As often, he was right, though congratulating himself too fulsomely. But it is not so clear that he had been right to urge Joseph to go to war against the Turks in 1783. Moreover, Kaunitz's hostility to any rapprochement with Prussia, a policy which tempted Joseph,[25] and his reluctance to seek English support became liabilities in 1789, if not before.

---

24 Ibid., 631.
25 Ibid., 382, 399–401, 557 and n.

Thomas Stamm-Kuhlmann

Gladstone: Morality in the Age of Popular Wars

Most of us have heroes – they may be heroes in history or heroes in scholarship. One of my *historical* heroes has always been Woodrow Wilson – one of my heroes *in scholarship* has always been Sigmund Freud. I was full of consternation when I learned that Freud detested Woodrow Wilson so deeply. In the introduction to the notorious book on Wilson that he has written together with William Bullitt, Freud says:

> *"I do not know how to avoid the conclusion that a man who is capable of taking the illusions of religion so literally and is so sure of a special personal intimacy with the Almighty is unfitted for relations with ordinary children of men"*[1].

In general, Freud's criticism of Wilson is based on the president's alleged hypocrisy, his rigour and the stubbornness that Wilson displayed already during his presidency of Princeton University.

At the age of sixteen, Wilson experienced a kind of awakening and was received as a grown-up member of the Presbyterian church. Around this time he put up a portrait of William Gladstone behind his desk and declared:

> *"That is Gladstone, the greatest statesman that ever lived. I intend to be a statesman, too"*[2].

This happened in 1873, during Gladstone's first term as prime minister. Obviously, the prime minister had already earned some fame across the Atlantic. The incident which had impressed young Woodrow Wilson may have been Gladstone's handling of the "Alabama" question, which both flattered the pride of the Americans and could be seen as a shining example of deference to legal claims in international affairs.

But I should begin in chronological order. One of the assumptions of our section is that Gladstone's political life fell into a period of "popular wars".

---

1 *Sigmund Freud*, *William C. Bullitt*, Thomas Woodrow Wilson, Twenty-eighth President of the United States. A Psychological Study, Boston 1966, XI.

2 Ibid., 15.

This meant that either governments had to see that their wars were made acceptable to the public or even that governments were driven by popular opinion into military adventures. In any case, moral reasoning had to take the masses into account as participants, at least passively, in the discourse.

This, again, is linked to certain developments in the constitution and in political culture, for which the rising importance of the press and of extra-parliamentary speeches may be an example.

Gladstone started his political career preoccupied with questions of domestic policy; some of his first concerns were of a rather esoteric kind such as his theological and ecclesiastical pursuits. In 1845, however, he became secretary of state for the colonies. The first of his famous appearances in debates on foreign policy occurred on 24 June 1850, when Gladstone was a simple M.P. The subject was the Don Pacifico incident, which had been used by Lord Palmerston as an occasion to emphasize Britain's prestige, saying that everybody who possessed British citizenship would be protected by the British government under any circumstances, even if, as in the case of Don Pacifico, the person was of Gibraltarian origin and making his living in Athens. To Gladstone, Palmerston's proud quotation of "civis Romanus sum" in this respect offered the opportunity to remind the world that Roman citizens had been treated as *members of a privileged class* and that the Romans had not recognized equal rights of other nations.

The first incident where Gladstone had to defend a major foreign policy decision from the position of a cabinet member we find when he was chancellor of the exchequer under the Earl of Aberdeen. In this position, Gladstone had to provide the financial means for the conduct of the Crimean War.

If there ever was a popular war in the 19th century, the Crimean War was it. A rather peace-loving prime minister, Aberdeen, was driven by the London "Times", although he suffered, in the words of the pacifist radical John Bright, "*as if every drop of blood that would be shed would rest upon his head*"[3]. In any case, fighting a war alongside with the Sultan against the Christian Emperor of Russia demanded at least a rationale other than solidarity among Christians. So the Prime Minister, Aberdeen, observed himself, saying that he did not know "*how he could bring himself to fight for the Turks?*"

3 *John Bright*, The Diaries, ed. by R. A. J. Walling, New York 1931, 164–165. Quoted from: *Akten zur Geschichte des Krimkrieges*, Serie III, Englische Akten, Vol.3, bearb. von Winfried Baumgart, München 1994, 34.

Gladstone replied,

> *"We were not fighting for the Turks, but we were warning Russia off the forbidden ground [...] We stand I said upon the ground that the Emperor has invaded countries not his own, inflicted wrong on Turkey & what I feel much more most cruel wrong on the wretched inhabitants of the Principalities."*

The inhabitants of Moldavia and Walachia were Christians; in Gladstone's words, they were the "*Christian subjects of the Sultan*", and against them, Gladstone said, he could not "*shoulder the musket*".

The main reason why the Czar was wrong, simply was that

> *"war had ensued and was raging with all its horrors: that we had procured for the Emperor an offer of honourable terms of peace which he had refused: that we were not going to extend the conflagration (but I had to correct myself as to the Baltic) but to apply more power for its extinction: and this I hoped in conjunction with all the Great Powers of Europe"*[4].

Obviously Britain's participation in the war did not cause any scruples for Gladstone. His opinion, however, soon changed. In February, 1855 he resigned from office because Lord Palmerston had accepted the motion for a committee of inquiry into the state of the army before Sebastopol. Gladstone regarded this as unconstitutional. He then remained a detached commentator of the events, although he was well aware that as a former cabinet member he bore responsibility for the fact that much harm and suffering had come about. While the government chose to call the outcome of the Vienna conferences of early 1855 a failure, Gladstone declared that these negotiations offered "*an admirable opportunity for putting an end to the dangers, horrors, and miseries of war*". Gladstone then went on, saying that from now on the war changed its character:

> *"The proposition I hold is this, that your war was just, that you have gained the objects of your war, and that if war was just while those objects were unattained, it becomes unjust if you continue to prosecute it after their attainment. I hold that you are now in danger of forfeiting and losing altogether the righteous and elevated character in which you have waged this war. [...] When the war broke out, by degrees our objects were considerably enlarged."*

4 *H. C. G. Matthew*, Gladstone 1809–1874, Oxford 1986, 85.

At the moment of the negotiations of Vienna, everything was agreed upon among the participants except one demand of the West, the limitation of the Russian fleet in the Black sea. This aim was qualified by Gladstone as

> "*inflict*[ing] *upon Russia indignity without taking away strength;* [...] *a standing insult to that Government – a standing deprivation of its own natural powers of defence – the first and most essential powers which belong to a government as such*".

During the negotiations of Vienna Russia had "*surrendered all her dangerous and inadmissible claims*" and therefore should be granted peace.

Gladstone then enumerated motives to continue the war, such as British prestige or the security of India, only to condemn them immediately.

> "*I frankly own that I shudder and tremble when I hear Gentlemen of humanity and high principle and character say that more blood must be shed – more human souls must be dismissed from the light of day* [...] *in order to satisfy the demands of military fame.*"

To continue fighting for success was termed by Gladstone as "*hideous, anti-Christian, immoral, inhuman*, a temptation of God, *guilty and unwise*". Then Gladstone went on to an argument of prudence which sounds rather Clausewitzian to me:

> "*Military operations are not the objects of war; they are its means and instruments*".

Gladstone underlined that Britain and France were not just two powers versus a third, but that they represented

> "*a principle* [...] *that which I may almost call a sort of European law which has prevailed* [...] *since the peace of 1815* [...] *to vindicate and enforce against* [Russia] *the public law of Europe*"[5].

As we see, Gladstone hardly left out any attribute that could be imagined. He appealed to everything that was great: law, mankind, and christianity, against anything that might sound like arguments of chauvinism and prestige. He also was outspoken and clear-cut in contrast to the "*language of ambiguity*" that, according to a resolution proposed by Disraeli[6], so far had been used by the government.

In his speech on the bombardment of the Chinese port of Canton of March 3, 1857, Gladstone again attacked the British double standard of judg-

5 Hansard, HC Deb 24 May 1855, vol. 138 cc 1037–1075.
6 Hansard, HC Deb 24 May 1855, vol. 138 cc 973.

ing Europeans and non-Europeans, which formed the core of the conflict that then had induced Sir John Bowring, the governor of Hong Kong, to order the attack on the Chinese harbour. This time, Gladstone offered to leave Christianity out of the debate, for he was able to appeal to something even more elementary:

> *"I will appeal to that which is older than Christianity, because it was in the world before Christianity – to that which is broader than Christianity, because it extends in the world beyond Christianity, – and to that which underlays Christianity, for Christianity in itself appeals to it – I appeal to that justice which binds man to man"*[7].

According to Gladstone, some "*subordinate officers of England, in a remote quarter of the globe, misconstrued the intentions of their country*". For him, the intentions of England could only be identical to the "*principles of right*". Therefore, it should have been the task of the executive to check those subordinates; since the government failed, the duty now fell upon the House of Commons.

> *"Every member of the House of Commons is proudly conscious that he belongs to an assembly which in its collective capacity is the paramount power of the State.* [...] *It is with us that it lies to determine whether this wrong shall remain unchecked and uncorrected* [...] *With you then, with us, with every one of us, it rests to show that this House, which is the first, the most ancient, and the noblest temple of freedom in the world, is also the temple of that everlasting justice without which freedom itself would be only a name or only a curse to mankind"*[8].

Here we find, first, a kind of natural right theory which was compatible with orthodox Christian beliefs but made it possible to reach freethinkers and believers of other religions as well, and, second, an argumentation which linked the cause of Britain directly to the cause of mankind.

Gladstone had spoken in support of a motion of censure by Richard Cobden which indeed was passed by a majority of 263 to 247. As Gladstone noted down in his diary, this was "*a division doing more honour to the h. of C. than any I ever remember*"[9].

At first glance, Gladstone's line of argumentation might be seen as hampering the British government's freedom of action. Under this absolute standard, there was no space left for arguments of raison d' état, for Realpolitik

---

7 Hansard, HC Deb 3 March 1857, exliv. 1798–1799.

8 Hansard, HC Deb 3 March 1857, exliv 1808.

9 *The Gladstone Diaries with Cabinet minutes and Prime-ministerial correspondence*, 14 vols., ed. by H. C. G. Matthew, Oxford 1978–1994, vol. V, 202.

or for the sacro egoismo della patria. On the other hand, by making the cause of Britain the cause of mankind, Gladstone accumulated a moral capital that Britain has drawn on in later conflicts; one might even say that there could not have been an Atlantic Charter without this precedent. Gladstone succeeded in an attempt that also has been made by German thinkers of the Napoleonic wars. Fichte and after him Marx had tried to make the German race the example of the human race and therefore the liberation of the Germans the prerequisite for the liberation of mankind. But this worked only within the constraints of German idealistic philosophy, while Gladstone referred to the traditions of classical philosophy that had found entrance even into the doctrine of the church of Rome. The German historian Rudolf Craemer, whose book *Gladstone als christlicher Staatsmann* came out in 1930, sees a conservative element in Gladstone's reasoning:

> "*The ius gentium is not just something abstract, but the concrete political form of the corpus Christianorum, now extended to the Chinese.*"[10]

While early on in his career Gladstone represented a rather marginal point of view which brought him in an alliance with the radical elements of the House, in his later years Gladstone mastered the task to gain popular support by addressing the people outside the walls of Parliament.

This opened a chance that popularity could be found not only on the side of jingoism and prestige, but that the masses could be won for the cause of universal justice that Gladstone had championed for so many years. As is said in the introduction of a new edition of the Midlothian speeches in 1970 by M. R. D. Foot,

> "*it was certainly not with any conscious idea of getting back to Downing Street that* [Gladstone] *went to Midlothian. He was impelled by a stronger, a simpler, and a more characteristic motive. He thought that Lord Beaconsfield's policies were wrong, and that it was his own duty to say so. This is the real core of the Midlothian speeches; and from this derives the principal impact that they made, both on their audiences in Scotland and – to a lesser, but still to a marked, extent – on their readers next day*"[11].

10 *Rudolf Craemer*, Gladstone als christlicher Staatsmann, Stuttgart/Berlin/Leipzig 1930, 242.

11 *W. E. Gladstone*, Midlothian speeches 1879 (The Victorian Library), with an introduction by M. R. D. Foot, Leicester 1971.

In the second Midlothian speech, we find the now famous lines which have been quoted in the House of Commons several times since 2001 – the lines on the savages of Afghanistan.

Especially addressing his female listeners, Gladstone said:

> *"I am not here before you as one of those who have ever professed to believe that the state which society has reached permits us to make a vow of universal peace, and of renouncing, in all cases, the alternative of war. But I am here to say that a long experience of life leads me* [...] *to a deeper and deeper conviction of the enormous mischiefs of war, even under the best and most favourable circumstances, and of the mischiefs indescribable and the guilt unredeemed of causeless and unnecessary wars."*

National pride, Gladstone went on, never could be a motive

> *"to blind you to the dictates of justice.* [...] *Remember the rights of the savage, as we call him. Remember that the happiness of his humble home, remember that the sanctity of life in the hill villages of Afghanistan among the winter snows, is as inviolable in the eye of Almighty God as can be your own. Remember that He who has united you together as human beings in the same flesh and blood, has bound you by the law of mutual love; that that mutual love is not limited by the shores of this island, is not limited by the boundaries of Christian civilisation; that it passes over the whole surface of the earth."*[12]

While this is a rather poetic expression of Gladstone's views, he shortly afterwards stated them in more doctrinary terms:

> *"Of all the principles, gentlemen, of foreign policy which I have enumerated, that to which I attach the greatest value is the principle of the equality of nations; because, without recognising that principle, there is no such thing as public right, and without public international right there is no instrument available for settling the transactions of mankind except material force."*[13]

Condemning Beaconsfield's imperialism was the major concern of the Midlothian speeches, with the result that Gladstone returned to Downing Street 10 in 1880. Again in office, he presided over one of the most significant steps

12 Ibid., 94.

13 Ibid., 123.

that the expansion of the empire has taken in the 19th century: the invasion of Egypt. The British and French control over the Suez canal and the treasury of Egypt, the unpopular rule of the Khedive, the loosening ties with the Ottoman Empire, all this had given rise to a native reform movement, led by Arabi Pasha, and finally to clashes between Arabi's followers and the British fleet. Alexandria was bombarded on July 11th, 1882. In a private letter to John Bright, who was at the time chancellor of the duchy of Lancaster and a long time companion of Gladstone's in the debates on the justification of wars, Gladstone on July 14th gathered his arguments to keep Bright from resigning in protest.

> "*The act of Tuesday* [the bombarding of Alexandria on the 11th] *was a solemn & painful one, for which I feel myself to be highly responsible* [...]
>
> *The general situation in Egypt had latterly become one in which everything was governed by sheer military violence. Every legitimate authority – the Khedive, the Sultan, the Notables, & best men of the country such as Cherif & Sultan Pachas had been put down, and a situation of* <u>*force*</u> *had been created, which could only be met by force. This being so we had laboured to the uttermost, almost alone but not without success, to secure that, if force were employed against the violence of Arabi, it should be the force of the Sovereign, authorised & restrained by the United Powers of Europe, who in such case represent the civilised world.*
>
> [...] *By this proceeding we have become responsible for the slaughter of many persons, defenders of the works, & probably for a few casualties in the town.* [...]
>
> *Such being the amount of our act, what has been its reception, & its effect?*
>
> *As to its reception, we have not received or heard of a word of disapproval from any Power great or small, or from any source having the slightest authority.*
>
> *As to its effect it has taught many lessons, struck a heavy perhaps a deadly blow at the reign of violence, brought again into the light the beginning of legitimate rule, shown the fanaticism of the East that massacre of Europeans is not likely to be perpetrated with impunity, & greatly advanced the Egyptian question towards a permanent & peaceable solution.*
>
> *I feel that in being party to this work, I have been a labourer in the cause of peace.*"

To all these arguments Bright replied that Gladstone's letter was

"so at variance with my views that they show my attempting to go with the Govt. is impossible"[14].

In another letter, to war secretary H. C. E. Childers, Gladstone wrote that he wished to avoid suspicions of invasion and therefore proposed to employ soldiers under civil authority to patrol the streets of Alexandria *"for purposes of order – and this is a case, & a very high case, of the order of the world"*. This idea proved to be impossible – the law demanded that soldiers were commanded by officers, as the war secretary said in his answer[15].

On July 31st, the Cabinet decided that the British expedition under Sir Garnet Wolseley had to *"put down Arabi & establish Khedive's power"*[16].

What had begun as a limited punitive action, ended up as the invasion of Egypt that Gladstone in the beginning wanted to avoid – but Gladstone was ready to produce a legal justification which ran as follows:

"When the Powers desired the armed interference of the Sultan as Sovereign it could not be had. The movements of the Sultan towards armed interference have been subsequent to the decision of England and to her proceeding to act upon that decision, and have taken no effect. Finally, the work has been accomplished not by the Proclamation of the Sultan but by the victory at Tel-el-kebir.

The former basis of the Sultan's power in Egypt has therefore failed both de jure and de facto."

Gladstone then intended Egypt to be neutralized and

"subject to all due provisions for the fulfilment of international engagements, it is presumed that England will make a firm stand for the reasonable development of self governing institutions in Egypt [...] *Would it not be well if this part of the subject were treated promptly with the Khedive, whose authority is presumed to be adequate for the purpose* [...]"

Gladstone proved

"Little sympathy could be expected from the Powers in promoting the development of securities for liberty; while in England they will be demanded, and will be hailed with satisfaction [...] *Under this head I*

14 *Gladstone Diaries* (note 9), vol. X, 298.
15 Ibid., vol.X, 299.
16 Ibid., vol.X, 306.

would suggest that the Khedive should signify in a formal manner his intention to use the aid of foreign officers with a view to the maintenance of Egyptian credit and the faithful discharge of obligations. But that the tenure of office by these functionaries should no longer be dependent on the consent of the respective Governments."[17]

What might look to an observer of today like one of the many imperialist adventures of the 19th century, to Gladstone had the highest moral backing he could dream of: consent of the powers, legalization by the formal approval of the Khedive, while the Sultan could be shown as de jure pushed aside – employment of force for the purpose of restoring the world order and policing a country of Eastern fanatics in the name of "civilization".

For such a success, Gladstone thought, it was appropriate to have the church bells rung and the guns fired in the London parks – which the secretary of war had failed to do.

While to Gladstone everything appeared logical and convincing, historians of later days were less enthusiastic. In censoring Gladstone's delay of help for the besieged Gordon at Khartoum, Gordon A. Craig wrote in 1990:

"Gladstone's fury at Gordon for what he called 'a war of conquest against people struggling to be free' [...] seemed to many to be a strange inconsistency, for if the Mahdi represented a people struggling to be free, why hadn't Colonel Arabi?"[18].

After all, Gladstone at least attempted to justify his handling of the Egyptian question vis-à-vis his old companions such as John Bright, but we hardly find any attempt to explain the annexation of Burma in 1886 to the world. In the cabinet notes of February 15, 1886, we read:

"Burmah: 1. Not to recede from the Queen's formal announcement."

In her Queen's speech, Victoria had mentioned "*the permanent incorporation of the Kingdom of Ava with my Empire*". Gladstone only stated that there was no prolonged deficit to expect from the annexation of Burma, and that there was "*no good native ruler*" available[19].

The following day, the Cabinet decided to recognize the annexation of Burma and to notify it to "*Foreign Powers*"[20].

---

17 Ibid., vol.X, 333.

18 *Gordon A. Craig*, Alexander L. George: Force and Statecraft. Diplomatic Problems of Our Time, 2nd ed., New York/Oxford 1990, 286 passim.

19 *Gladstone Diaries* (note 9), vol.XI, Oxford 1990, 497.

20 Ibid., vol.XI, 498.

If we attempt to find what might be considered as the essence of Gladstone's reasoning throughout the decades, I find it most striking that Gladstone always aims at assembling as much authority as possible on his side. Much authority can be derived from universal principles. Since they apply to every comparable situation on any spot on the planet and any part of mankind – including the hill tribes of Afghanistan – they have the backing of humanity. They are as absolutely binding as any law of mathematics – as categorical as the categorical imperative, although I have not found so far any reference to Kant's philosophy in his texts. If humanity is not sufficient, there can always be the appeal to God, as we have seen in the speech on the Crimean war. On earth, the proper expression of the universalism of principle is the unity of powers. Gladstone in the first of the Midlothian speeches of 1879 connected the word *authority* to the united action of the powers in the Eastern question:

> *"Let coercion be applied by the united authority of Europe – that is to say, for it is not an exaggeration so to put it, by the united authority of the civilised world applicable to this case."*[21]

Gladstone even devoted time to the question whether America forms an element of this united authority. He answered in the affirmative but preferred to omit America for practical reasons. In the 19th century, America was not yet concerned by the affairs of the Balkans. In any case, there is more to this unity than the combined military strength of Britain, France, Germany, Austria and Russia. There is the coercion of the mind that derives from the fact that what everybody agrees on, must be true. We find here still the influence of the quotation from St. Augustine that Gladstone puzzled over in his younger years when he was confronted with Manning's and Newman's conversion to the church of Rome: "*Securus iudicat orbis terrarum.*"

There are two possibilities why someone might prefer this style of thinking. One reason might be a logical mind. Certainly Gladstone was gifted in that respect. He had a decent formal education in mathematics, philosophy and rhetorics – the best one could get at the time. But Gladstone was no rationalist. As an evangelical, he believed in the revealed truths of Christianity which more than once demand from us the sacrificium intellectus. He was convinced that there was a personal God who intervened in human affairs. Therefore, he might have been ready to accept unique events, exceptions, and surprising turns in the course of history. But he preferred an under-

21 *Midlothian speeches* (note 11), 54.

standing of the world according to few principles which were applicable to every issue everywhere.

So we get to the second possible explanation. The authority connected to these principles is a great help – if you have to deal with a sensitive conscience. Gladstone, with his habitual introspection and the scourge ready at hand, led a life in permanent respect of his huge super-ego. In this situation, his universal principles gave him security.

What can frighten a man who feels in harmony with his conscience, the principles, mankind and even God? Once assembled on the side of one individual, these awesome allies give him a feeling of great strength and righteousness. A person governed by these feelings might behave "tactlessly" and "self-righteously", disavowing even his own cabinet colleagues as Gladstone did in 1870, when he wrote his famous un-anonymous article on the Franco Prussian war[22]. From here, we are not far from bigotry, and we might understand why a person convinced of his rectitude might develop enormous stubbornness and even a tendency to overlook the stains on his own clean record.

What are we supposed to think when we learn that Gladstone, at the time when he authorized the bombardment of Alexandria and the invasion of Egypt, had no less than 37 per cent of his portfolio invested in Egyptian Tribute Loans? In the Commons he said on 12 July:

> "*Undoubtedly it is not for the exclusive or special interest of the bondholders of Egypt – and indeed it is almost wholly without any reference to them – that the proceeding of yesterday* [i.e. the bombardment] *was taken.*"[23]

In fact the price of the Tribute Loan stock "*rose not on the bombardment of Alexandria, but on Wolseley's expedition in August with its implication of at least a short-run occupation,*" as can be shown by the contemporary press. The Loan Stock of 1871,

> "*bought at 42, [...] had fallen to 57 in the summer of 1882; but by December 18182 it was worth 82.*" "*That is, Gladstone's 1871 stock was worth £ 17.100 before the invasion of Egypt and 24.600 soon after it.*"[24]

22 *Derek Beales*, Gladstone and His First Ministry, The Historical Journal XXVI (1983), 997.

23 *H. C. G. Matthew*, Gladstone, 1875–1898, Oxford 1995, 136, quoting from Hansard 272.173.

24 *Matthew* (note 23), 135.

Certainly there was no awareness of the peculiarities of insider deals as we know it today. However, Gladstone recorded his moral scruples in his diaries, for example when he had been reading pornography, but the diaries remain silent on such miraculous blendings of financial interest and the "*authority of the civilized world*".

Already in his lifetime observers said that Gladstone had a "dual character". Toward the end of Gladstone's life John Robertson, the radical journalist, put it this way:

> "*It is this perceived combination, as it were, of rectitude and crookedness in Mr. Gladstone's mental processes, the consistence in him of admitted moral elevation with a curious moral versatility, which looks like levity of principle and conviction – it is this that makes his character such a theme of dispute. If we can explain that duality, we shall understand and explain Mr. Gladstone.*"[25]

25 Quoted from: *Travis L. Crosby*, The Two Mr. Gladstones. A Study in Psychology and History, New Haven/London 1997, 4.

Karina Urbach

Bismarck: Ein Amateur in Uniform?

Woran erkennt man Bismarck? In Karikaturen dienen drei Dinge als Schlüsselreize: ein bulliger Mann, eine Uniform und Kürassierstiefel. Bismarck ist in das kollektive Gedächtnis als die Personifikation eines Kriegers eingegangen. Seine Taten wurden jahrzehntelang in populärwissenschaftlichen Biografien mit einer martialischen Sprache umschrieben: Neben den Einigungskriegen ‚kämpfte' er den Kulturkampf, den Kampf gegen die Sozialisten, die Polen, Hamburger Geschäftsleute und seine bevorzugte bête noir – die Ehefrauen der Hohenzollern.

Der Reichskanzler selbst sah sich verständlicherweise als multiplere Persönlichkeit. Wenn Faust von sich glaube, zwei Seelen in seiner Brust zu haben, so meinte Bismarck, könne er dies leicht überbieten – in seiner Brust würden endlos viele Seelen miteinander kämpfen: „*Es geht zu wie in einer Republik.*"[1] Eine militärische Seele gehörte jedoch lange Zeit nicht in diese Republik. Bismarck entdeckte den Soldaten in sich erst spät. Sein Schlüsselerlebnis war der Krieg von 1870.

Mehrere Historikergenerationen haben diesen Krieg als Beispiel für den klassischen Machtkampf zwischen Militär und Politik beschrieben[2]. Einige entdeckten hier sogar die Wurzeln für das Versagen der deutschen Regierung im Ersten Weltkrieg. Bismarck – so die These – habe schon 1870 nur unter größter Kraftanstrengung den Primat der Politik gegen die Militärs durchsetzen können.

Im Folgenden soll gezeigt werden, dass diese Argumente auf einer einseitigen, Bismarck-zentrischen Sichtweise beruhen. Die Rollenverteilung im deutsch-französischen Krieg war tatsächlich sehr viel komplexer und kann nicht auf den einfachen Gegensatz Militär-Politik beschränkt werden.

---

1 Zitiert u.a. in: *Karina Urbach*, Between Saviour and Villain: 100 Years of Bismarck biographies, in: The Historical Journal, Vol. 41: 4 December 1998, S. 1141.

2 Grundlegend: *Gerhard Ritter*, Staatskunst und Kriegshandwerk. Das Problem des ‚Militarismus' in Deutschland, Bd. 4, München 1968.

Sowohl Generalfeldmarschall Moltkes als auch König Wilhelms Verhalten müssen in dieser Gleichung ein stärkeres Gewicht erhalten. In einem ersten Schritt soll gezeigt werden, dass Bismarck im Krieg von 1870–71 nicht – wie er später behauptete – allein die Rolle des weisen Staatsmannes spielte, sondern sich massiv in militärische Entscheidungen einmischte und dadurch eine Kettenreaktion freisetzte.

Otto von Bismarck entdeckte die Liebe zum Militär erst spät. Für seine Generation war der Beruf des Soldaten alles andere als attraktiv gewesen und der junge Bismarck entschied sich daher bewusst gegen diese Laufbahn. Er versuchte sogar die Militärpflicht zu vermeiden und trat 1838 nur widerwillig bei den Potsdamer Gardejägern ein. Die Armee vergaß und vergab nicht. Zwanzig Jahre später musste Bismarck seine gesamte Überzeugungskunst aufwenden, um eine standesgemäße Uniform tragen zu dürfen. 1859 war er zum preußischen Gesandten in St. Petersburg ernannt worden und argumentierte, dass man an diesem Hof ohne Uniform gesellschaftlich nicht überleben könnte. General von Manteuffel zögerte trotzdem lange ihm die Erlaubnis hierfür zu erteilen[3]. Erst 1862 durfte Bismarck bei einer Parade eine Kürassieruniform mit dem Charakter eines Majors vorführen[4] und nach dem erfolgreichen Krieg gegen Österreich 1866 wurde er zum Generalmajor ernannt. Von nun an trug er ununterbrochen seine Uniform – für einen Politiker eine eher ungewöhnliche Entscheidung. Sie hatte durchaus ein psychologisches Element. Bismarck wollte König Wilhelms militärischer Welt näherkommen, die ihm ansonsten fremd war. Wie sehr er sich nach der Aufnahme in diese besondere Welt sehnte, zeigt unter anderem ein Weihnachtsbrief an den Kaiser, darin bedauerte Bismarck, den Hohenzollern nur vom Schreibtisch aus und nicht auf dem Schlachtfeld gedient zu haben[5].

Bismarcks Liebe zur Armee ist insofern erstaunlich, da er von dieser wenig Gegenliebe erfuhr. Wie in allen Armeen der Welt existierten auch in der preußischen eine Vielzahl von Exklusionsmechanismen, die sich schon allein an der Sprache zeigten – ‚Zivilist', ‚civvis' oder ‚les pekins' waren Schimpfworte. In den Augen der Armeeführung blieb Bismarck bei aller Mimikry ein Amateur in Uniform. Lothar Gall hat beschrieben wie man im General-

3 Denkwürdigkeiten aus dem Leben des General-Feldmarschalls Kriegsministers Grafen von Roon, Breslau 1892, Bd. 2, S. 86.

4 Ibid.

5 Zitiert in: *Otto Pflanze*, Bismarck and the Development of Germany: The Period of Unification, 1815–1871, Princeton 1971, S. 43.

stab den ‚Zivilist im Kürassierrock' verspottete[6]. 1870 wurde dies für Bismarck zu einer ernsthaften Belastung.

In den Kriegen von 1864 und 1866 hatte er kaum Probleme bei der Zusammenarbeit mit den Militärs gehabt. In seinen ‚Gedanken und Erinnerungen' beschreibt der Reichskanzler, wie sich im Sommer 1870 alles veränderte:

> *„Schon bei der Abreise in Köln erfuhr ich durch einen Zufall, daß beim Ausbruch des Krieges der Plan festgestellt war, mich von den militärischen Beratungen auszuschließen.* [...] *Ich konnte das aus einem Gespräch des Generals von Podbielski mit dem Grafen Roon entnehmen, dessen unfreiwilliger Ohrenzeuge ich wurde.* [Podbielski] *äußerte laut seine Befriedigung: ‚Diesmal ist also dafür gesorgt, daß uns dergleichen nicht wieder passiert.'"*[7]

‚Nicht wieder' habe sich – so vermutete Bismarck – auf den Krieg von 1866 bezogen, als er unter anderem auf die Änderung der Marschrichtung von Wien nach Preßburg bestanden habe. Als ‚Strafe' hierfür wäre er 1870 von den Militärs gemieden und bei Einquartierungen sowie bei der Naturalverpflegung außergewöhnlich schlecht behandelt worden (als kulinarischer Kenner war Letzteres schmerzlich für ihn).

Inwieweit Bismarck jedoch tatsächlich ein derartiges Gespräch zwischen Roon und Podbielski belauscht hat, ist schwer zu verifizieren. Beide waren zum Zeitpunkt der Publikation von ‚Gedanken und Erinnerungen' bereits tot. In seiner Autobiografie konnte Bismarck daher ungestraft über die Behandlung durch die Generalität lamentieren:

> *„Der Generalstab [war] der Ansicht gewesen [ich käme] erst wieder zu Wort, wenn die Heeresleitung die Zeit gekommen findet, den Janustempel zu schließen."*

Sie habe dabei verkannt, dass das doppelte Gesicht des Janus bereits darauf hinweisen würde, dass

> *„die Regierung eines kriegsführenden Staates auch nach anderen Richtungen zu sehen habe als nach dem Kriegsschauplatze"*[8].

Bismarck stellte sich in seinen Erinnerungen also als den weisen, aber einsamen Politiker dar, der von einer Schar dumpfer Soldaten umgeben war. Während die Armee in einem ethnozentrischen Tunnelblick verharrte, dachte er allein in internationalen Zusammenhängen. Da die militärische

---

6 Zitiert in: *Lothar Gall*, Bismarck. Der weiße Revolutionär, Berlin 1980, S. 442.
7 *Otto von Bismarck*, Gedanken und Erinnerungen, Neuauflage Berlin 1998, S. 345.
8 Ibid.

Führungsspitze ihn gemieden habe, hätte er viel Zeit im *Hotel des Reservoirs* verbringen müssen, wo man die zweite Liga untergebracht hatte. Er wäre dort unter anderem gezwungen gewesen den Times-Korrespondent nach militärischen Neuigkeiten zu fragen[9].

Das Bild eines schlecht genährten Bismarcks, der in einer Hotellobby ‚herumlungert', um Informationen zu erhaschen, wirkt nicht besonders überzeugend. Der Bundeskanzler gehörte nicht zu der Sorte Mensch, der lange Zeit freiwillig in den antechambers der Macht verweilte. Tatsächlich beschloss Bismarck früh, einen ‚Konkurrenzhof' aufzubauen und selbst eine militärische Rolle zu spielen. Nach zwei erfolgreichen Kriegen sah er sich als ernst zunehmenden Militärexperten. Schon am 17. August 1870 schrieb er an seine Frau:

> „*Die Führung der ersten und zweiten Armee ist ungeschickt im Mißbrauch der todesmuthigen Tapferkeit unserer Leute, nur Faust, ohne Kopf, und doch siegen wir.*"[10]

Bismarcks Chance, sich als Militär zu beweisen, kam, als die militärische Situation zunehmend eskalierte. Der Krieg hatte im Herbst 1870 eine völlig neue, unbekannte Dimension angenommen. Es war ein nationaler Krieg mit hohem emotionalem Einsatz geworden, aber vor allem verlief er nicht planmäßig. Er drohte sich in die Länge zu ziehen, die Zahl der Toten schnellte in ungeahnte Höhen und dies wiederum steigerte die Verbitterung auf beiden Seiten. 1864 und 1866 waren begrenzbare, schnelle Kriege gewesen. Politiker wie Militärs bevorzugen verständlicherweise derartige Feldzüge. 1870 gab es jedoch mehrere Gründe, die den Krieg verlängerten: Die Gefangennahme Napoleons III. am 2. September 1870 hätte unter normalen Umständen den Krieg beendet, doch für die Franzosen war es ein Volkskrieg geworden, ein ‚*guerre à outrance*' – der von nun an auf Guerillataktiken basierte. Damit hatte auf deutscher Seite niemand gerechnet. Die Führungsschicht wurde unruhig.

Man muss kein Experte im Krisenmanagement sein, um zu verstehen, dass eine Führung, die unter enormen Zeit- und Erwartungsdruck steht, Gefahr läuft auseinanderzubrechen. Auf deutscher Seite wussten alle Beteiligten, dass der Krieg so schnell wie möglich beendet werden musste, da sonst die Großmächte einschreiten würden[11]. Klaus Hildebrand hat gezeigt, wie energisch der Krieg von Gladstone und der englischen Öffentlichkeit

9 Ibid., S. 347.

10 Bismarck an seine Frau, zitiert in: *Bismarck Briefe*, hg. v. Herbert Rothfels, Göttingen 1955, S. 359.

11 *Eberhard Kolb*, Der Weg aus dem Krieg: Bismarcks Politik im Krieg und die Friedensanbahnung 1870/71, München 1990.

verurteilt wurde[12]. Darüber hinaus galt es eine drohende Intervention Rußlands abzuwenden.

Kriege gebären Gerüchte, Verschwörungstheorien und vor allem Schuldzuweisungen. Auch in der sicheren Etappe kann ein Krieg sich bewusstseinsverändernd auswirken. Bei Tischgesprächen, die von Moritz Busch protokolliert wurden, zeigte Bismarck nun eine martialische Seite[13]. Seine Sprache schien unter dem Druck der Ereignisse die Form des Kasinos anzunehmen. Unterstützt wurde er hierbei von seiner Frau Johanna, die unter anderem forderte *„die Gottlosen* [Franzosen] *sollen ausgerottet werden*“[14] – eine Ausnahme solle höchstens für französische Kinder gemacht werden. Derart animiert schlüpfte Bismarck in die Rolle des Landsknechts und forderte die größtmögliche Brutalität gegenüber dem Feind: Man solle Geiseln nehmen und Terror gegen die Zivilbevölkerung ausüben. Ausfälle dieser Art würden noch ein Nachspiel haben. Moritz Buschs Aufzeichnungen ‚Bismarcks große Tage‘ wurden in Auszügen später in der „Times“ veröffentlicht und beschädigte das Bild des klugen, humanen Bismarck in Großbritannien schwer[15]. Ideengeber für die brutale ‚Kriegstaktik‘ war ein neuer Tischgast des Bundeskanzlers: Der amerikanische General Philip Sheridan. Sheridan hatte während des amerikanischen Bürgerkrieges feindliche Dörfer niedergebrannt und empfahl diese Methode ausdrücklich[16]. Die Begeisterung Bismarcks hierfür deutet nicht auf einen Staatsmann hin, der langfristig darauf bedacht sein musste, eine friedliche Nachkriegsordnung anzustreben. ‚Authentische‘ Militärs wie Moltke und Wilhelm I. lehnten derartige Vorschläge daher ab. Doch die Anspannung dieser Tage führte bei Bismarck dazu, überall Verschwörungen zu wittern. In seinen Tischgesprächen wetterte er unter anderem, dass die Armee die Bombardierung von Paris nur ab-

---

12 *Klaus Hildebrand*, “No Intervention”: Die Pax Britannica und Preußen 1865/66–1869/70: eine Untersuchung zur englischen Weltpolitik im 19. Jahrhundert, Oldenbourg 1997. Siehe zu Gladstones Außenpolitik Thomas Stamm-Kuhlmanns Beitrag in diesem Band und ders., Bismarck und Gladstone, Friedrichsruher Beiträge, Band 37, Friedrichsruh 2008.

13 Graf Bismarck und seine Leute während des Krieges mit Frankreich. Nach Tagebuchblätter von Dr. Moritz Busch. 1878. *Ders.*, Bismarcks große Tage. Ein Chronist erlebt die Reichsgründung, Holzminden 1990 (Neudruck). Erschienen in englischer Übersetzung als Our Chancellor. Sketches for a Historical Picture by Moritz Busch, London 1884.

14 Ibid., S. 147.

15 Siehe hierzu *Karina Urbach*, Portrait of a Giant: Otto von Bismarck im zeitgenössischen Urteil Großbritanniens, in: *Otto von Bismarck im Spiegel Europas*, hg. v. Klaus Hildebrand und Eberhard Kolb, Paderborn 2006, S. 178.

16 *On the Road to Total War: The American. Civil War and the German Wars of Unification, 1861–1871*, hg. v. Stig Förster und Jörg Nagler, New York 1997, S. 129.

lehne, weil der Kronprinz mit einer Engländerin verheiratet sei und Königin Victoria bekanntermaßen Paris schützen wolle. Dies klang nicht nach dem sonst so rational agierenden Außenpolitiker. Besonders Moltke war entsetzt über den unbekannten Soldaten Bismarck. Aus seiner Sicht hatte der Bundeskanzler von Anfang an drei Aufgaben zu erfüllen gehabt: erstens, einen Kriegsgrund zu finden; zweitens, den Krieg zu finanzieren und drittens, die Nichteinmischung der Großmächte zu garantieren[17]. Stattdessen stellte der hochnervöse Bundeskanzler ununterbrochen Forderungen an die Armee[18]. Moltke argumentierte, diesen Forderungen nicht folgen zu müssen – ein Generalstabschef unterstand nicht einem Politiker. 1871 bestand er daher in einem Brief an den König auf einer endgültigen Klärung der Verhältnisse:

> *„Ich glaube, daß es gut sein würde, mein Verhältnis zum Bundeskanzler definitiv festzustellen. Bisher habe ich dasselbe dahin aufgefaßt, daß der Chef des Generalstabes – besonders im Krieg – und der Bundeskanzler zwei gleich berechtigte und einander unabhängige Behörden unter Eurer königlichen Majestät directem Befehl sind, welche sich gegenseitig in Kenntnis zu halten haben.“*[19]

In Moltkes Augen waren sie also Ebenbürtige.

Nach Clausewitz sind die Sphären Politik und Militär getrennt: Der Plan zu einem Krieg ist eine politische Entscheidung; der Feldzug selbst ist eine militärisch-technische Entscheidung. Bismarck und Moltke hatten beide Clausewitz gelesen, wenn auch im Falle Bismarcks nicht eindeutig ist zu welchem Zeitpunkt[20]. Etwas gelesen zu haben bedeutet bekanntermaßen nicht, dass man es verstanden oder akzeptiert hat. Beide – Bismarck wie Moltke – interpretierten Clausewitz in ihrem Sinne: Moltke kannte ‚Vom Kriege‘ und benutzte Clausewitzsche Terminologie für seinen Aufsatz ‚Über Strategie‘, den er Mitte 1871 veröffentlichte. Hierin argumentierte er, dass zwar die Politik den Krieg für ihre Zwecke benutzten sollte, dass aber die Armee am Erfolgreichsten während des Krieges agieren könne, wenn sie unabhängig wäre.

Dass das Militär sich der Politik unterordnen muss, war eine Forderung, die erst mit Clausewitz in das öffentliche Bewusstsein gelangte. Samuel Finer

---

17 Siehe hierzu *Moltke. Vom Kabinettskrieg zum Volkskrieg. Eine Werkauswahl*, hg. v. Stig Förster, Bonn 1992.

18 *On the Road to Total War* (Anm 18), S. 135.

19 Zitiert in: *Beatrice Heuser*, Clausewitz lesen! Eine Einführung, München 2005, S. 73.

20 Beatrice Heuser glaubt Bismarck habe Clausewitz nie gelesen, Ibid, S. 72. Sie beruft sich hierbei auf: *Jehuda Wallach*, The dogma of the battle of annihilation. The theories of Clausewitz. Contribution in Military Studies 45, London 1986. Diese Bismarck-Aussage ist jedoch nicht zu verifizieren.

hat in seinem Buch ‚The Man on Horseback' länder- und epochenübergreifend die Rolle von Militärs untersucht[21]. Drei Faktoren geben der Armee eine eindeutige Überlegenheit gegenüber den Politikern: erstens, die Armee hat eine perfekte, allen anderen Institutionen überlegene Organisationsbasis; zweitens, sie hat einen großen emotional-symbolischen Status; drittens, sie hat das Monopol auf die Waffen.

Finer stellte daher die Frage, warum – bei all dieser Überlegenheit – Militärführer nicht öfter gegen politische Einmischungen rebellieren. Überspitzt gefragt, warum sollten sie überhaupt Politikern gehorchen[22]? Diese Frage stellt sich natürlich vor allem im Kriegsfall. In einem Krieg sind Politiker abhängig von militärischen Erfolgen. Moltke hatte daher einen enormen Vorteil gegenüber Bismarck. Die 1860er Jahre waren ein gutes Jahrzehnt für preußische Generale gewesen. Die Armee erlangte großes Prestige und siegreiche Soldaten werden naturgemäß geliebt. Moltkes Machtbasis war folglich groß, er hätte jederzeit mit einer Art Coup drohen können.

Der Grund, warum Moltke am Ende Kompromisse machte und keinen totalen Krieg gegen Frankreich führte, war nicht, dass Bismarck ihn durch sein angeblich politisches Geschick davon abhielt. Der Grund war ein anderer: Tatsächlich handelte es sich bei den Kämpfen in der preußischen Führungsspitze um eine Dreiecksgeschichte, wobei die dritte Person in dieser Beziehung häufig vernachlässigt wird. Zwei Männer – Bismarck und Moltke – buhlten um einen dritten Mann: König Wilhelm.

Wilhelm spielt in der Forschung immer eine etwas schwache, zu vernachlässigende Rolle. In Wirklichkeit ist er aber 1870 die entscheidende Figur in dieser Auseinandersetzung. Er besitzt etwas, was die beiden anderen Mitspieler nicht haben: Legitimität.

Weder Bismarck noch Moltke hätten jemals den Krieg diktatorisch führen können, da ihre Herrschaft keine Legitimität aufwies. Als Monarch war Wilhelm der oberste Kriegsherr. Damit war er während der ganzen Auseinandersetzung der einzige und entscheidende ‚Arbiter'.

Ein wichtiger Grund, warum dieser Aspekt in der Forschung übersehen wird, ist die Tatsache, dass demokratisch denkende Historiker unterschätzen, welche enorme Wirkung die monarchische Legitimität *unterbewusst* für alle Beteiligten hatte. Sowohl Bismarck wie Moltke fühlten sich eng mit dem Haus Hohenzollern verbunden. Die preußische Armee hatte traditionell eine besondere Beziehung zu ihrem Herrscherhaus und der preußische

21 *Samuel Finer*, The Man on Horseback: The Role of the Military in Politics, London 1962.

22 Ibid, S. 5.

'Beamte' Bismarck verglich sich gerne mit Strafford, der für seinen König sterben würde[23].

Ein weiterer Grund für die Vernachlässigung der Rolle Wilhelms I. ist, dass Historiker zwar alle Dokumente administrativer Provenienz – und natürlich einen Großteil der Privatbriefe – besitzen aber die vielen mündlichen Beschwichtigungen und Vermittlungsversuche Wilhelms nur teilweise rekonstruieren können. Hinter den Kulissen betrieb der König eine geschickte Personalpolitik. Bereits am 3. September 1870 appellierte er bei einem gemeinsamen Dinner an Militärs und Politiker:

> *„Sie Kriegsminister Roon, haben unser Schwert geschärft; Sie General von Moltke, haben es geleitet, und Sie, Graf Bismarck, haben seit Jahren durch die Leitung der Politik Preußen auf seinen Höhepunkt gebracht."*[24]

Moltke sah sich zwar als Diener des Königs – aber nicht als Diener der gerade im Amt befindlichen Regierung. Er unterwarf sich am Ende also der Entscheidung seines Souveräns, nicht der Anordnung eines Politikers. Dies ist ein wichtiger Unterschied. Es ist aber auch ein Unterschied, der im 20. Jahrhundert *gefährlich* werden konnte. Wenn Armeen sich von der jeweiligen Regierung distanzieren, können sie Gefahr laufen ihre eigenen Vorstellungen davon zu entwickeln, was im Interesse 'ihres Landes' ist.[25] Militärs haben system- und länderübergreifend immer wieder versucht diesen Unterschied zu machen.

Im deutsch-französischen Krieg war der König ein entscheidendes Regulativ. In dem gesamten Konflikt ging es um die Frage, welche Argumente ihn letztendlich überzeugten. Moltke stand dem König mentalitätsmäßig näher als sein Rivale Bismarck. Der Bundeskanzler blieb bei dieser Auseinandersetzung trotzdem im Vorteil. Er konnte bekanntermaßen die stärksten Charaktere brechen – mit seiner großen Argumentationskraft, aber auch mit seinem enormen Charme. Wilhelm war klug genug, dies zu erkennen. Bismarck, so der König, könne einfach *„alle Nachrichten in* 'couleur de rose' *verwandeln"*[26]. Natürlich hatte der Bundeskanzler hierfür auch eine perfekte Infrastruktur zu Verfügung – einen Geldfonds zur Pressebeeinflussung. Diese 'passende' Berichterstattung half Bismarck auch, seine Ideen

23 Der 1st Earl of Strafford (1593–1641) opferte sich für seinen Herren, König Charles I.

24 Zitiert in: *Hanns Martin Elster*, Graf Albrecht von Roon. Sein Leben und Wirken, Berlin 1938, S. 500.

25 Finer (Anm. 23), S. 23.

26 Urbach (Anm. 1), S. 1154.

beim König durchzusetzen. Wenn die Zeitungen folglich die Bombardierung von Paris forderten, dann hatte dies letztendlich auch Auswirkungen auf Wilhelms Entscheidungsfindung[27].

Ein Krieg ist das klassische Drama in drei Akten: Disposition, Katharsis auf dem Schlachtfeld, Triumphzug. Die Szenenbilder können variieren, doch das letzte Bild ist entscheidend, um die Opfer zu rechtfertigen: Der Triumphzug von 1871 war eine derart perfekte Inszenierung. Bismarck und Moltke ritten einträchtig nebeneinander her. Das Ideal der Männerfreundschaft schien erfüllt. Die Diadochenkämpfe blieben der Öffentlichkeit verborgen. In einer Monarchie wie Preußen stand natürlich 1871 vor allem Wilhelm I. im Mittelpunkt der Feierlichkeiten. Im Berliner Opernhaus wurde am 17. Juni 1871 eine Vorstellung von lebenden Bildern gegeben, in denen der Krieg bereits historisiert wurde: An dem alten Barbarossa ziehen die großen Momente der deutschen Geschichte vorbei: 1189 die Kreuzfahrer, 1678 der große Kurfürst, 1760 der alte Fritz, eine Szene aus den Freiheitskriegen, Kriegsbilder von 1864 und 1866. Krönender Abschluss war das Jahr 1870: Die Germania wird von preußischen, bayerischen und württembergischen Soldaten getragen[28].

Kriege enden jedoch nur auf der Bühne mit einem Triumphzug, in der Realität gehen sie in den Köpfen der Beteiligten weiter – häufig über Generationen hinweg. Auch Historiker fechten Kriege bekanntermaßen immer wieder neu aus. Für das Ausland war es von Anfang an Bismarck, der diesen Krieg zum Erfolg geführt hatte[29]. Er unterstützte diese Wahrnehmung nach Kräften. Auch im Deutschen Reich wurde er – durch geschickte Eigenpropaganda – noch zu Lebzeiten zum Sinnbild der Einigungskriege. In Karikaturen zeigte man ihn von nun an als die deutsche Führergestalt: der Deichhauptmann, der Lotse[30] – immer ein Mann in Uniform.

---

27 Schon Friedrich der Große hatte die preußischen Zeitungen mit seinen Kriegsberichten versorgt. Bismarck ging nun einen Schritt weiter und weitete die ‚inspirierte' Presseberichterstattung auf das Ausland aus. Siehe hierzu auch *Frank Becker*, Bilder von Krieg und Nation, Oldenburg 2001, S. 43. *Richard Kohnen*, Pressepolitik des Deutschen Bundes. Methoden staatlicher Pressepolitik nach der Revolution von 1848, Tübingen 1995, S. 135f.

28 Siehe hierzu die Tagebucheintragung von Baronin Spitzemberg, 17.6.1871, Das Tagebuch der Baronin Spitzemberg. Aufzeichnungen aus der Hofgesellschaft des Hohenzollernreichs, Göttingen 1989, 127.

29 Urbach, (Anm. 17), S. 171.

30 *Rolf Parr*, ‚Zwei Seelen wohnen ach in meiner Brust'. Strukturen und Funktionen der Mythisierung Bismarcks 1860–1918, München 1992, 6.

Zusammenfassend muss man über den Staatsmann Bismarck und seine Beteiligung im deutsch-französischen Krieg zwei Anmerkungen machen, die nicht in dieses perfekte Bild passen. Erstens: Es herrscht in der Forschung heute eine Bismarck-zentrische Sichtweise vor, wenn es um die Beurteilungen des 70/71 Krieges geht. Am Ende entschied über militärische und politische Angelegenheiten Wilhelm I. Die martialischen Tischgespräche zeigen, dass Bismarck nur ein großer Staatsmann werden konnte, weil er eben *nicht* an der Spitze des Staates stand.

Zweitens: Die Rollen Politik (Bismarck) gegen Militär (Moltke) waren 1870–71 nicht eindeutig festgelegt. Bismarck agierte in dieser Situation nicht nur als Politiker. Er wollte in der Tradition eines Friedrich des Großen beides sein: Politiker und Militär. Moltke überschritt ebenfalls seine Kompetenzen. Er hatte sich hierbei nicht zum ersten Mal in das politische Revier eingemischt. Schon 1867 drang er darauf, den Krieg gegen Frankreich zu beginnen. Beide, Bismarck und Moltke, hatten also einen gefährlichen Rollentausch betrieben – und Clausewitz ignoriert.

Der Konflikt Militär gegen Politik existiert zeit- und länderübergreifend. Auseinandersetzungen und Überlappungen der beiden Einflussfelder sind in einem Krieg eher die Norm. Schlussfolgerungen für einen deutschen Sonderweg sollten aus diesem Beispiel also nicht gezogen werden. Bismarck und die Militärs würden sich in den folgenden Jahren noch viele Auseinandersetzungen liefern[31]. Die Armee würden dabei selten gegen den Reichskanzler gewinnen – vor allem nicht bei der entscheidenden Frage eines Präventivkrieges. Bismarck hatte aus seiner kurzen Episode als ‚Landsknecht' viel gelernt.

31 Siehe hierzu Michael Schmid, Der Eiserne Kanzler und die Generäle. Deutsche Rüstungspolitik in der Ära Bismarck (1871–1890), Paderborn 2003.

Keith Robbins

The Welsh Wizard who won the War: David Lloyd George as War Leader

The United Kingdom of Great Britain and Ireland which went to war in 1914, it scarcely needs to be said, was a limited monarchy and so it remained at its conclusion. A politician, whatever his dynamism or charisma, could never entirely occupy centre-stage. Men died 'For King and Country', not for David Lloyd George. It was the King, Queen and the Prince of Wales, after war had been proclaimed, who appeared on the balcony of Buckingham Palace in August 1914. It is appropriate, therefore, to begin with a reminder, in the case of Britain, that 'leadership' in wartime was a shared function of monarch and minister, though that is not to say that it took the same form or required the same capacities.

The 44-year old George V came to the throne in May 1910. He had been educated and trained as a naval officer. Had it not been for the death of his elder brother in 1892 – whose fiancée he inherited – he would have continued in the career he loved. Perhaps the only consolation for his role as heir to the throne had been that it afforded him more time to devote to his massive stamp collection. His skill with guns had to be restricted to their use on his Sandringham estate. Essentially a private man, he did not share his father's enthusiasm for high living or for energetic and sometimes fraught mingling with European royalty[1]. After George became Prince of Wales in 1901, his father had kept him reasonably well-informed about affairs of state. He knew what lay ahead of him. He was Commander-in-Chief of the Armed Forces. He was well aware, however, that the time had long since passed when, in person, the monarch was expected to lead his men into battle. In the event of a war, the role he would play would be as symbol of national unity. The Crown would be above parties.

---

1 *Keith Robbins*, The Monarch's Concept of Foreign Policy: Victoria and Edward VII in: *An Anglo-German Dialogue: The Munich Lectures on the History of International Relations*, ed. by Adolf M. Birke, Magnus Brechtgen and Alaric Searle, Munich 2000, 125–129.

George V was therefore a cipher, but still more than a cipher. His training as a naval officer, even though it had not taken place in the *Dreadnought* era that had now arrived, had left him with insight into the minds and ways of sailors and soldiers. He would not be ill at ease in the roles and expectations placed upon him in wartime. He was, however, beholden to ministers. Ministers, in turn, would nevertheless have to listen to whatever advice he might offer. He was also accessible to generals and admirals, which would perhaps mean that he would be lobbied by them. As it happens, he came to the throne at a moment of constitutional crisis when conventional understandings of the relationship between the monarch and the government were being put to the test. It still rested with the Crown to grant or withhold a request to dissolve parliament or to form a government. He had immediately been placed in a quandary by his Prime Minister, H. H. Asquith, whose government had gained a renewed, though reduced, majority at the general election in January 1910. The issue was what should be done with the House of Lords which had broken with the convention that it should not tamper with a money bill which had been passed by the House of Commons. A majority of peers had taken the view in November 1909 that the budget sent for approval was far more than a simple money bill. After inconclusive negotiations, the Prime Minister cornered a reluctant but eventually acquiescent monarch into agreeing to create a sufficient number of peerages to ensure the passage of the budget. Another election was held in December, after which the measure was passed. It did not prove necessary to create new peerages. In August 1911 the Parliament Act ended the power of veto possessed by the Lords and replaced it by a power to delay.

It was widely supposed that the man who was really behind this crisis and the constitutional change with which it concluded was the man who had framed the 1909 budget, the man who was 'the People's Champion': David Lloyd George[2]. Initially appointed President of the Board of Trade on the formation of the Liberal government, he had succeeded Asquith as Chancellor of the Exchequer in 1908. His budget was not merely a tinkering with taxation. It was a social landmark – the dream of Liberalism for generations realised at last, as he put it. As A. J. P. Taylor argued, the things he did were all his own, "*things not envisaged by ordinary liberalism or by the party programme*"[3]. He had attacked dukes with a relish which upset both Buckingham Palace and the Prime Minister's wife. She did not like a reference to lords being high 'like cheese'. Such attacks, she thought, lost rather than gained votes. Lloyd George was unrepentant. After a visit to Balmoral

2 *B. K. Murray*, The People's Budget, 1909/10: Lloyd George and Liberal Politics, Oxford 1980.

3 *A. J. P. Taylor*, Lloyd George: Rise and Fall, Cambridge 1961, 18–19.

HANS-VON-HERWARTH-PREIS

Georgie Williams

Prof. Dr. Dieter Weiß

1. Vorsitzender der Prinz-Albert-Gesellschaft

Coburg, 1. Oktober 2010

in September 1911 he found the new monarch "*a very very small man*". All the King's sympathy, he wrote, was with the rich and he had very little pity for the poor. Such an observation, however, did not lead the Chancellor of the Exchequer to decline the chunk of venison from a deer shot by his monarch and thoughtfully sent to North Wales for his consumption. The King and Queen, he wrote privately, were "*hostile to the bone*" to all who were working to lift the workmen out of the mire. Court life had no appeal to Lloyd George. Its whole atmosphere, he said, reeked with Toryism. It was true that people were very civil to him, but so they would be to a dangerous wild animal[4]. To some contemporaries, of course, Lloyd George himself was a dangerous wild animal.

It was not certain, moreover, that he could be tamed. There was nobody in the contemporary British political elite who was quite like him. To some commentators, he was mesmerizingly mysterious. He had come from 'nowhere', that is to say from North-West Wales. Eighteen months younger than his 'very very small monarch', he was proud to belong to a very small nation. He had been elected to the Commons in 1890 as MP for the Caernarfon boroughs when he was merely 27. After the death of his schoolmaster father when he was a baby, he had been brought up by his mother and shoemaker uncle in Llanystumdwy, a small village unknown to, and unpronounceable by, the English public. His ambition was burning. He would not stay a small town attorney, but make his career in the wide world. He spoke to his constituents in Welsh and English. In the House of Commons, of course, only English was deployed. The chapel oratory in which he had been reared had left its mark. He was a Nonconformist, though a nonconforming one. He liked singing hymns. He had a half-English half French/Italian secretary who became his mistress. He had a wife largely at home in Wales. Duality, in short, seemed to sum him up. He had a London life and a Welsh life. Only Churchill used his first name 'David'. He was generally referred to as 'L. G'. The return of a native to his constituency was like the return of a God. Messages relaying the progress of his train along the North Wales coast were relayed to audiences rapturously awaiting his arrival. People wanted to categorize him but he slipped through their labels. Later, J. M. Keynes famously (or infamously) in his *Essays in Biography* described him as a "*goat-footed bard*" who was "*rooted in nothing*". This "*half-human*" visitor had stumbled into the twentieth century from the enchanted woods of Celtic antiquity. He was a medium and a vampire rolled into one[5]. Of course, it helped, in making such an assessment if, like

4 *Lloyd George Family Letters 1885–1936*, ed. by Kenneth O. Morgan, Cardiff and London 1973, 154–155.

5 *J. M. Keynes*, Essays in Biography, London 1961, 35–36.

Keynes, one had been educated at Eton and King's College, Cambridge – where men believed themselves to be fully human.

It would certainly be safe to assume that Keynes knew more about economics than the Chancellor of the Exchequer. But did that matter? Running a department, being in a government, was not a matter of knowledge. It was a matter of managing, perhaps manipulating men. Contemporaries, before the war, observed that Lloyd George was supremely good at getting 'the sense of the meeting', whether private or public, at 'getting the drift', at grasping what people felt though did not say. Indeed, some suspected that it was not so much the substance of what was decided that fascinated so much as the process. One did not need, like the Prime Minister, H. H. Asquith, to have gained First Class Honours in Classical 'Mods' and 'Greats' – Latin and Greek History, Literature and Philosophy – at Balliol College, Oxford, to see what had to be done in politics and, even more important, to see how to do it. Civil servants never ceased to be staggered by the skill and apparent ease with which this 'uneducated' Welshman constructed convincing speeches from the scraps of pertinent information that they passed to him.

Such qualities and capacities very largely manifested themselves in the disputes and battles of domestic politics. Asquith was some ten years older than Lloyd George. Gladstonian precedents suggested that Liberal Prime Ministers liked to last a long time. It might take exceptional and unanticipated circumstances to give Lloyd George an opportunity to reach 10 Downing Street. It might take a war, a major war. Any such speculation – and there was always speculation – ran into difficulty. What was Lloyd George's attitude to war and what did that attitude say about his capacity to be a 'war leader'? Here, too, there was ambiguity surrounding him. There was a certain view that Welsh Nonconformists were 'pacifists' – a newly-coined and itself ambiguous term. Certainly, one of the most famous of their nineteenth-century figures, Henry Richard, had been secretary of the Peace Society and had come to be called 'The Apostle of Peace'. Historians now generally agree, however, that in fact Welsh people differed very little in their attitudes to war from their English or Scottish contemporaries[6].

It therefore comes as no great surprise that in his late teens, far away from the politics of Westminster, Lloyd George was expressing satisfaction when the British bombarded Alexandria in 1882 (under a Liberal government).

---

6 *Keith Robbins*, A Welshman and the Pursuit of Peace: Henry Richard and the Path to the 1850 Frankfurt Peace Congress, in: *Internationale Beziehungen im 19. und 20. Jahrhundert: Festschrift für Winfried Baumgart*, ed. by Wolfgang Elz and Sönke Neitzel, Paderborn 2003, 19–35; *Goronwy J. Jones*, Wales and the Quest for Peace, Cardiff 1969; *Wales and War: Society, Politics and Religion in the Nineteenth and Twentieth Centuries*, ed. by Matthew Cragoe and Chris Williams, Cardiff 2007.

It is true, later, that he was categorized as a 'Pro-Boer' during the Anglo-South African War of 1899–1902 and, at first sight, this might identify him as a 'pacifist' and anti-imperialist. He was neither. He was not hostile to the British Empire as such. His opposition to the South African War stemmed not from any supposed identification with the Afrikaners – as a member of another bullied 'small nation' – as from a belief that engineering a war had been clumsy and unnecessary. It was a stupid war rather than a morally culpable one. At the Treasury, in general, he supported the increasing naval expenditure of these years. His interest in foreign policy grew and his public statements and interventions sought to make it clear, in particular instances, that where her vital interests were at stake Britain would not be treated as if it were of no account.. He told his brother that his speech on this theme at the Mansion House during the Agadir crisis had "*done the trick*"[7]. A foreign secretary plagued on occasion in 1911 and subsequently by Liberal critics of his foreign policy did not find in Lloyd George a 'disloyal' colleague[8]. Lloyd George wrote to Grey in September 1911 that war was

> "*so much in the reckoning as to render it urgently necessary for us to take every step which would render the issue of war more favourable, always provided that such a step does not include the chance of precipitating war*"[9].

Lloyd George himself took pleasure in the assertion of Dr. E. J. Dillon, a well-known newspaper foreign correspondent, that he had taken Gladstone's place on the continent of Europe and had become the best known name in all countries. If one wants to put it so, these were the years of the 'education of a statesman' in which he learnt something of the nature of international diplomacy and of the precarious balance between peace and war in pre-1914 Europe. His speeches outside his departmental brief could sometimes cause disquiet elsewhere in the government, but not generally so. They could be taken to indicate that Lloyd George, as time passed, was 'maturing' as a politician. Even so, one might say that he was always careful to hedge his bets. When the Anglo-German Friendship Society eventually held its long-planned conference in November 1912 Lloyd George was careful to attend the dinner held in connection with it in the House of Commons[10].

---

7 *K. M. Wilson*, The Agadir Crisis, the Mansion House Speech, and the Double-Edgedness of Agreements, Historical Journal 15 (1972), 513–532.

8 *Keith Robbins*, Sir Edward Grey: A Biography of Lord Grey of Fallodon, London 1971, 290–291.

9 *David Stevenson*, Armaments and the Coming of War: Europe 1904–1914, Oxford 1996, 190.

10 *Paul Laity*, The British Peace Movement 1870–1914, Oxford 2001, 189.

If minds turned to leadership in war, it was to the whizz-kid ex-Tory, Winston Churchill, that minds turned. Here was a product of Sandhurst, soldier, war correspondent, descendant of the great Duke of Marlborough, running the Admiralty in 1911 at the age of 46. He had seen war at first hand, albeit in a colonial context. He had visited the battlefield at Blenheim. He attended German army manoeuvres on more than one occasion. The Kaiser, with whom he exchanged a few words, referred to him by his first name. Churchill, therefore, had studied, and was studying, war in a way that Lloyd George never appeared to do. If there was a war, there was clearly, in Winston, a recognizable war leader to hand. The 'pro-Boer' label still hung around Lloyd George. Surely, he, and a good section of the Liberal party with him, would be likely to oppose British intervention in a European war?

Indeed, Lloyd George did not relish or welcome the prospect. There was, it seems, a real prospect that if the pattern of events in the summer of 1914 had followed even a slightly different courser he would have opposed British intervention, at least pro tem. On 3 August he described himself as moving through a nightmare world. He claimed that he had so far kept the Cabinet out of war but if a small nationality like Belgium were attacked by Germany all his traditions and prejudices would be engaged on the side of war[11]. Dissenter though he was, in the event he did not join the small band of dissentients[12]. The violation of Belgian neutrality gave him a ready identification with the plight of small nations. He appealed to British 'Good Samaritans' to help Belgium "*that little country bleeding on the roadside*". Speaking to Nonconformist and London Welsh audiences, who might have been expected to be of a generally 'pacific' if not 'pacifist' disposition, he was able to be convincing. If Lloyd George supported intervention, his audiences were prepared to believe, then the cause must be justified. It strengthened the position of a minister, in a war government, if 'the people' had no perception of him as a bellicose militarist. Put another way, he could guarantee, as perhaps no one else could, national unity so long as the war lasted. His popular standing meant that, to some extent at least, he could distance himself from the understandable impetus towards 'closing ranks' and accepting, where the conduct of war was concerned, hallowed procedures and such 'regime constraints' against which he had successfully kicked in domestic affairs.

*

No one, sitting at the Cabinet table in August 1914 could judge how long the war would last or precisely what its character would be. It appeared likely,

11 *Lloyd George Family Letters* (note 4), 167.

12 *Cameron Hazlehurst*, Politicians at War: July 1914 to May 1915: A Prologue to the Triumph of Lloyd George, London 1971, 54–65.

however, that it would strain the machinery of government and raise again the always familiar issues of the relationship between civilians and military in the planning and execution of warfare. No one, probably not even Lloyd George, would have believed that he would be, in 1918, 'the man who won the war'. Around the table at the beginning were men of different stamp, politicians all, certainly, but none of them tested by the particular demands of war. Was the job of Foreign Secretary, for example, in abeyance? Was it 'business as usual' or was it 'all change'? Politicians who thrived in peacetime might flounder in war. Personality did matter, not because a wizard could wave a wand, but because only an individual could pull together all the discordant elements inescapably present in a 'war machine' and impose a sense of direction. It required self-confidence in the face of expertise. It was not that such capacities were not required in peacetime as that the demands were more pressing and the consequences of failure more evident. War had to be waged at so many different levels. Did it not require such a streamlining of procedures as to render obsolete conventions of collective decision-making? Moreover, as it went on, leadership was not only a matter of meetings and the making of a host of small decisions. It was a question of sustaining national morale and keeping 'the people' on side. That could still only be a matter of speech-making and printed text. It was not until some years after the war that Lloyd George ever used a microphone when speaking in public. It did no harm that the country had a Prime Minister who could have been a star of the music hall. His ability to handle interventions from his audiences had not yet deserted him. It had to be a 'popular' war. The country was more 'democratic' in its form than it had been in any other major conflict. A leader who could not keep "the people" behind him could not win the war. Radicals were already arguing strongly that it was time

> *"for the extension principles of popular representative government to a region that is still dominated by a spirit which is alien, if not positively inimical, to the spirit of modern democracy"*[13].

Such questions, and other analogous ones, bring us back to Lloyd George. The British government which ended the war in November 1918 was very different from the government which went to war in August 1914. Lloyd George was at the heart of the transitions that occurred. We can never, of course, satisfactorily answer the counter-factual question of what would have happened if he had remained demurely in office at the Treasury until the end of the conflict (though admittedly the thought of him remaining

---

13 *Arthur Ponsonby*, Democracy and Diplomacy, London 1915, 9; *Marvin Swartz*, The Union of Democratic Control in British Politics during the First World War, Oxford 1971.

demurely anywhere is difficult). It is possible that an Asquith would have 'lost the war' if he had remained in office longer. So might a rugged technocrat, a master of planning, but somebody who was without charisma (as we might now put it). Lloyd George had a confidence that he could do the job. Both friends and sufficient erstwhile opponents came to the same conclusion. That was why, eventually, the position came his way in December 1916. It was precisely because, in the manner which has been described above, he was an 'outsider' that he could transcend the established networks of British politics.

It had looked, initially, in 1914, as though things would go on "*as normal*". The government remained a Liberal government. There was no rush for parties to come together "*in the national interest*". Lloyd George remained as Chancellor of the Exchequer. What was significant, however, was the appointment as Secretary of State for War of that imperial soldier, Lord Kitchener of Khartoum. Did this mean that the civilians were 'handing over' the war to the soldiers? To some extent , in the raising of his 'New Armies' and in directing strategy, it looked initially as though this might be the case. By May 1915, however, it has been argued that it was becoming obvious that wars between industrial societies would necessarily involve whole societies. Britain could not continue to fight the war successfully by waging less than total war. It was Lloyd George pre-eminently who saw that the only road to victory lay through total war[14]. It was then that the first major governmental change took place. A coalition government was formed, with Asquith still as Prime Minister. Lloyd George left the Treasury to become the new Minister of Munitions. One key development with which he was closely associated was the introduction of compulsory military service for men of appropriate age – a step opposed by a substantial body of Liberal opinion. It was quintessentially a wartime post. It was just the kind of role which suited his admitted and admired skills. He had to identify demand, remove bottlenecks, and improve supply. The task was not 'ideological'. However, the story of the steps taken to organise labour for the munitions factories, Lloyd George wrote in his memoirs, could not be completed without paying tribute to the work of the King. The words uttered 'man to man' by the Head of State, on his many visits to factories, according to Lloyd George, ran like wildfire through the works. It was by such involvement, it was claimed, George V established himself "*more firmly than ever in the affection of his people*"[15]. In exercising leadership in war, a monarch could do things a politician could not do, and vice versa.

14 *David French*, British Economic and Strategic Planning 1905–1915, London 1982, 173

15 *David Lloyd George*, War Memoirs, London 1934, 191–192.

When Kitchener was drowned in June 1916 Asquith, after some delay, appointed Lloyd George to succeed him as Secretary of State for War. It was now evidently a post which civilians not soldiers should occupy. Lloyd George's energy was already being favourably compared, in some quarters, with that of the Prime Minister. He could get on with the Conservative ministers who now helped to make up the government. Six months later, he was Prime Minister (though precisely how this happened need not detain us). It appeared that he was the man who could administer the "knock-out blow" which would win the war. Peace advocates, however, speculated about what he might do when he realized, as he surely would, that such a blow could not be delivered.

> *"I fancy he is clever enough to try to arrange peace even though he may be bellowing ever so loudly about organisation for war. Does he want peace do you think? I think he does."*

So wrote Morgan Jones in December 1916, a fellow-Welshman and future Labour MP but then a conscientious objector in detention. Lloyd George, thought Bertrand Russell, was not worse than the rest of mankind. Indeed, he belonged to the best ten per cent. It was just that the human race as a whole was vile[16].

Ignorant of these speculations on what he might or might not do, Lloyd George proceeded to form a small War Cabinet, the members of which were charged with very specific responsibilities. The appointment to it of Lord Milner was one example of the way in which Lloyd George enlisted men of proven capacity with little regard to previous animosity or party background[17]. It was the Prime Minister's own personality which kept this diverse show on the road. Nor did his enthusiasm for innovation stop with the making of fresh ministerial appointments. He devised for himself at 10 Downing Street a team of assistants or advisers who came to be dubbed 'the garden suburb' (a reference to their accommodation). By such means he could, if he thought necessary, by-pass formal civil service advice. This was administrative 'total war'. Naturally, those who found themselves on occasion demoted were not happy with the Prime Minister's evident lack of respect for convention and tradition.

---

16 *Keith Robbins*, The Abolition of War: The 'Peace Movement' in Britain 1914–1919, Cardiff 1976, 113–117.

17 *John A. Turner*, The Formation of Lloyd George's 'Garden Suburb': 'Fabian-like Milnerite Infiltration'?, Historical Journal 20 (1977), 165–184; *War and the State: The Transformation of British Government 1914–1919*, ed. by Kathleen Burk, London 1982.

There was no doubt, however, that the Prime Minister was in charge, for better or worse. There were those, not unexpectedly, who thought that these changes were for the worse. Among them were generals. "*Lloyd George and the Generals*"[18] is the title of at least one book. The Prime Minister naturally features in the biographies of every major British wartime general. The detailed ramifications of their disputes and disagreements cannot be pursued here and an overall verdict is not easily made. The disputes over strategy, most conspicuously the long-running arguments between 'Easterners' and 'Westerners' were fundamental. Lloyd George wanted to create what he called 'a ring of flame' in South-East Europe. It was necessary for the Allies to develop a co-ordinated strategy for theatres other than the Western Front[19]. It was scarcely surprising that there should be such arguments. Behind them, however, lay stereotypes and perceptions, on both sides, which ran deep through British social life. "*Haig does not care how many men he loses*", Lloyd George told Frances Stevenson his mistress/secretary in January 1917,

> "*He just squanders the lives of these boys ... He seems to think they are his property. I am their trustee*"[20].

In dealing with Haig and other generals Lloyd George left little room for doubt that he considered them second-rate men. Such verdicts may be unjust but it was the expression of a view which he felt deeply. It is not surprising that, in turn, disparaging remarks are to be found in return. Early in the war, finding its political direction chaotic, Douglas Haig remarked that a Dictator was wanted, but the difficulty was to find one. Much later, Haig can be found complaining that Lloyd George seemed to resent his bulking larger in the public eyes at all. The Prime Minister wished to shine alone[21]. There were moments when Lloyd George almost ceased to shine at all. In the crisis that led in February 1918 to the resignation of the Chief of the Imperial General Staff, Sir William Robertson, he almost fell from power. He was also vulnerable in the following May in a debate arising from disclosures in the press by General Maurice concerning the number of men serving on the western front. In short, he never became 'the Dictator' for whom Haig

---

18 *David R. Woodward*, Lloyd George and the Generals, Newark/London/Toronto 1983.

19 *Keith Robbins*, Politicians, Diplomacy and War in Modern British History, London 1994, 215–238; *Michael G. Fry*, Lloyd George and Foreign Policy, Montreal/London 1977, 267–286,

20 *Lloyd George. A Diary by Frances Stevenson*, ed. by A. J. P. Taylor, London 1971, 139.

21 *Gerard De Groot*, Douglas Haig 1861–1928, London 1988, 197 and 297.

had apparently at one stage been looking. One historian argues that Lloyd George's imagination, flexibility and boldness, qualities which served him well in the ordinary run of politics, outran themselves. Attracted by bold schemes, he could be insufficiently interested in the logistical difficulties in their implementation which soldiers obstinately insisted upon[22].

'Personality', Lloyd George was to find, cannot resolve all issues and problems at a stroke. The Prime Minister was where he was because of personality but he was not sufficiently buttressed by the party loyalties and those institutional affiliations and allegiances which transcend the person[23]. In 1915 he wrote a preface to *Through Terror to Triumph*, a collection of his wartime speeches thus far. In it he promised that if the country's preparations were characterised by grip, resolution, and a prompt readiness in every sphere, victory was assured[24]. The 'People's Champion' had certainly provided grip, but it had not become the grip of a wartime dictator. He could not take major decisions without the agreement of at least some of his colleagues. He spurned seductive suggestions that there could be peace by negotiation. He could sustain the popular will to 'fight to the finish' as perhaps no one else at the time could[25]. He could use language which suggested that this would be the war to end war. At its conclusion, he asserted that

> *"a great attempt must be made to establish by some international organization an alternative to war as a means of settling international disputes"*[26].

Constitutional proprieties were honoured. The king, blossoming into the House of Windsor, remained on the throne. There had been much terror by 1918 and the triumph, when it came, proved personally to be politically short lived. Too much personality, as his friend and rival Winston Churchill was also to find at the conclusion of another war, can grate. A. J. P. Taylor argued that, whether consciously or not, Lloyd George had aimed at becoming the sole voice of the people by destroying the existing parties and reducing politics to a collection of individuals. If so, it did not work. The wizard whose wand had once administered the 'knock-out blow' was reduced to being a spectator at the ringside[27].

---

22 *David R. Woodward* (note 18), 335.

23 *Kenneth O. Morgan*, Lloyd George's Premiership: A Study in Prime Ministerial Government, Historical Journal 13 (1970), 130–157.

24 *David Lloyd George*, Through Terror to Triumph, London 1915, ix.

25 *V. H. Rothwell*, British War Aims and Peace Diplomacy 1914–1918, Oxford 1971.

26 *George W. Egerton*, Great Britain and the Creation of the League of Nations, Chapel Hill 1978, 61.

27 *Peter Clarke*, A Question of Leadership: From Gladstone to Thatcher, London 1992, 83–105.

Christopher Clark

Supreme Warlord: Kaiser Wilhelm II in Wartime

To examine the last four years of the reign of the last German Kaiser under the rubric of leadership in wartime might at first glance appear an ironic, or even sarcastic enterprise. Contemporaries and historians alike agreed and agree that Wilhelm II entirely failed to play the leadership role implied in the term 'Supreme Warlord' (*Oberster Kriegsherr*), widely used from around 1900 to denote his person. Yet a closer look at the Kaiser's role in decision-making suggests that this controversial monarch could and did exercise a certain influence on policy choices at crucial junctures in the wartime history of the Kaiserreich. There can be no question, of course, of the Kaiser's manifesting 'statesmanly' virtues under the pressure of an increasingly bitter and ultimately doomed war effort. But the relationship between the Kaiser's failure as a wartime 'leader' and his readiness to deploy the residue of power still reserved to him under the Imperial Constitution of 1871 is worth exploring, not only for what it reveals about the legal and structural complexities of the *Kaiserreich*, but also for the sometimes surprising light it sheds on the Kaiser himself.

*

The Prussian-German monarch was – in theory at least – a military commander, and from the beginning of his reign, Wilhelm saw this aspect of his responsibilities as a crucial dimension of his sovereignty. Article 63 of the Imperial constitution of 1871 stipulated that the 'unified army' of Germany, comprising the 'entire territorial strength of the Empire' stood in peacetime and in war "*under the command of the Kaiser*"[1]. In a Cabinet Order issued just a few weeks after his accession to the throne, Wilhelm announced the creation of a new military establishment to be known as the 'Headquarters of His Majesty the Kaiser and King'. By contrast with the traditional

1 Die Verfassung des Deutschen Reiches vom 16. April 1871, accessed online at: http://www.dhm.de/lemo/html/dokumente/verfassungkai/index.html, last accessed 28 March 2010

military entourage, whose members attended the monarch as rep- es of their respective branches of the service, the generals and ad- ants of the 'Headquarters' were personally selected by the Kaiser, with the assistance of the chief of the military Cabinet[2]. Throughout the pre-war era, Wilhelm took a keen interest in armaments issues and in the drafting of military and naval legislation and even aspired to play a leading part in the annual army manoeuvres. From around 1900, these pretensions were reinforced by the increasingly widespread use – warmly welcomed and encouraged by the Kaiser himself – of the non-constitutional term 'Supreme Warlord' to describe the imperial sovereign[3].

The monarch's determination to preserve and extend the extra-parliamentary royal *Kommandogewalt* had important consequences for the political and constitutional history of the Reich. It also affected the structural evolution of the higher military and naval commands. Above all, it militated against the emergence of a central body capable of coordinating the activities of the respective services and reconciling their priorities. During the reign of Wilhelm I, responsibility for personnel, training, armaments and deployment was gradually parcelled out to a range of competing agencies: the General Staff, the Ministry of War, the Military Cabinet, and so on. Wilhelm II made no attempt to reverse this state of affairs, indeed he further atomized the command structure by stepping up the number of military and naval command posts that reported directly to the emperor[4]. This was part of a conscious strategy to create an environment that would permit the untrammeled exercise of the monarchical command function. "*After deciding to exercise personally the supreme command over My navy, just as over My army*", Wilhelm declared in a cabinet order of 14 March 1899,

> "*I do not consider it practical that there should stand between myself and the various leaders a central institution of command, which would merely have the purpose of conveying my orders*"[5].

In reality, however, Wilhelm's capacity to exercise a command function was narrowly circumscribed. His participation in army manoeuvres proved

2 *W. Deist*, Kaiser Wilhelm II in the context of his military and naval entourage, in *Wilhelm II: New Interpretations*, ed. By J. C. G. Röhl and N. Sombart, Cambridge 1982, 169–92; here 182–3.

3 *W. Deist*, Kaiser Wilhelm II als Oberster Kriegsherr, in *Der Ort Kaiser Wilhelms II in der deutschen Geschichte*, ed. By J. C. G. Röhl, Munich 1991, 25–42; here 26; on the military dimension of Wilhelm's sovereignty more generally, see *E. Fehrenbach*, Wandlungen des deutschen Kaisergedankens, Munich 1969, 122–4, 170–2.

4 *Deist* (note 3), 30; *Deist* (note 2), 176–8.

5 *E. R. Huber*, Heer und Staat in der deutschen Geschichte, 2nd ed., Hamburg 1938, 358.

highly disruptive, partly because Wilhelm was a poor tactician and partly because the Chief of the General Staff (CGS) Count Alfred von Schlieffen took the view that if the Kaiser participated then he must not be permitted to lose in battle simulations: "*as Kaiser he cannot be beaten by one of his generals*" – a princple that rendered military exercises largely useless[6]. Schlieffen's successor as CGS, Helmuth von Moltke, made Wilhelm's exclusion from further involvement a condition of his taking office in 1906.

Wilhelm also lacked an overview of strategic planning. He was apprised of the broad drift of General Staff strategy for the eventuality of war but he was not informed of the details of military planning in the immediate pre-war years, possibly because the secrecy-obsessed General Staff regarded him as a potential security leak. Moreover, neither the military and naval cabinets, nor the imperial 'Headquarters' possessed the resources to support the monarch in a genuine command function. Wilhelm was thus unable to perform the kind of coordinating role that would have compensated for Germany's relative lack of unitary command institutions comparable with the Conseil Supérieure de la Guerre in France and the Committee of Imperial Defence in Great Britain. Even in the last years before the outbreak of war, virtually no concrete preparations were made for coordinated military-naval operations, and there was no attempt to dovetail the strategies cooked up by the military planners with the options pursued by German diplomacy. Germany remained, in this sense, "*strategically leaderless*"[7].

In the sphere of military command, then, as in so many other aspects of the emperor's role, there was a yawning gap between rhetoric and reality. Wilhelm is reported to have declared that he would be "*his own CGS in time of war*"[8]. If he did indeed make this claim, the outbreak of war in August 1914 cured him of such delusions; he lost no time in formally transferring to the CGS the right to issue operational orders in his name. The Staff Chief thereby became "*the man who is actually in control of the supreme army command*"[9]. Early in August, Wilhelm promised that he would abstain from interfering in the running of operations and, to the surprise of the military commanders, he stuck to this promise throughout the war.

---

6 Diary entry of 23 July 1904, in *H. v. Moltke*, Erinnerungen, Briefe, Dokumente 1877–1916, ed. by E. v. Moltke, Stuttgart 1922, 296.

7 *Holger Afflerbach*, Kaiser Wilhelm II. als Oberster Kriegsherr im Ersten Weltkrieg. Quellen aus der militärischen Umgebung des Kaisers 1914–1918, Oldenbourg 2005, 20.

8 *W. Groener*, Das Testament des Grafen Schlieffen. Operative Studien über den Weltkrieg, Berlin 1930, 79.

9 *E.v. Falkenhayn*, General Headquarters and Its Critical Decisions 1914–1916, London 1919, 4.

Moreover, it soon became clear that the Kaiser lacked not merely the strategic expertise and vision, but also the psychological resources to sustain him in a wartime leadership role. The outbreak of war appears to have cast him into a state of nervous exhaustion that lasted – with brief interruptions – as long as the war itself. When the former chancellor Bernhard von Bülow visited the Kaiser at the beginning of August 1914 (!), he noted the emperor's "*terrified, distorted countenance*". The mood swings and oscillations that had always been a signal feature of this monarch's personality became steadily more pronounced. Wilhelm could be moved to bloodthirsty exultation by rumours of success, but was also easily cast down and prone to bouts of defeatism. His moods soared and plunged in response to the rapidly changing news from the front. On 6 September 1916, for example, it was reported that the Kaiser "*looked very bad*" because he had just received an alarming report on the condition of the First Guards Division on the Somme. But he apparently spent the very next day in a "*highly elevated mood*", following reports of the fall of the Rumanian fortress of Tutrakan[10].

The Kaiser thus increasingly became a peripheral figure, a "*prisoner of his generals*" as an Austro-Hungarian minister later recalled[11]. He stayed at the *Großes Hauptquartier* throughout the war years, but complained of not being informed or consulted by the decision-makers[12].

> "*If people in Germany think I lead the army*", he remarked in November 1914, "*then they are very mistaken. I drink tea and cut wood and go walking, and then I hear from time to time that this and that has been done, just as the gentlemen wish … The only one who is a bit nicer to me is the head of the Military Railways Department – he tells me everything his doing or planning.*"[13]

"*I interfere as little as possible*", he told the Prussian Minister of War Adolf Wild von Hohenborn in the summer of 1915.

> "*After all, Falkenhayn has to maintain the fiction that I personally issue all the orders.*"[14]

10 *Afflerbach* (note 7), 39

11 *R. Chickering*, Imperial Germany and the Great War 1914–1918, Cambridge 1998, 33; *L. Cecil*, Wilhelm II, 2 vols., Chapel Hill 1989, 1996, vol. 2, Emperor and Exile, 1900–1941, 209.

12 *H. Afflerbach*, Falkenhayn: Politisches Denken und Handeln im Kaiserreich, Munich 1994, 236.

13 Cited in *Fehrenbach* (note 3), 216.

14 Adolf Wild von Hohenborn, letter to his wife of 7 June 1915 in: *id.*, Briefe und Tagebuchaufzeichnungen des preußischen Generals als Kriegsminister und Truppenführer im Ersten Weltkrieg, ed. By Helmut Reinhold and Gerhard Granier, Boppard am Rhein 1986, 64.

Visitors to the Imperial Headquarters were often struck by the mood of unreality they encountered there as the emperor sat dining on the silver service of Frederick the Great while treating his guests to vivid third-hand tales of German feats in battle: "*piles of corpses six feet high* [...] *a sergeant killed twenty-seven Frenchmen with forty-five bullets*", and so on[15]. There were embarrassing visits to the front. After a visit by the Kaiser to the 11th Reserve Corps in June 1916, for example, the corps staff chief recalled:

> "*His Majesty looked well, was gracious and spoke of banalities. What he said about the war was such that is best to remain silent on the matter. The commanding general, who is such an imperial loyalist, was ashen-faced.*"[16]

To an army commander who had just fought off a French attack with heavy losses, the Kaiser issued the following advice in the autumn of 1915:

> "*Just order your men to fix bayonets and throw those rascals out again.*"[17]

The cabinet chiefs whose task it was to attend the Kaiser during the war years complained (privately) of the claustrophobia and tedium of evenings spent listening to imperial monologues, humoring his latest enthusiasms (die philological reconstruction of the Hittite language was in vogue during the summer of 1916), taking part in conversations that never seemed to gain any real momentum, or playing cards to pass the time[18].

Historians have thus rightly stressed the role of the war in displacing the emperor from the centre of affairs. As Lamar Cecil has put it:

> "*the once-vaunted persönliches Regiment, with its mighty throne and ruler, became in wartime a backseat occupied by a neglected, ill-informed and increasingly inconsequential figurehead*"[19].

In John Röhl's justly celebrated three-volume biography of the Kaiser, the chapters on the role played by the 'Supreme Warlord' during the World War merit considerably fewer pages – only 69 out of a total of 3,747 densely printed sides – than any other phase of the Kaiser's childhood, youth or adult life before the abdication and flight to Holland. And the reason for this uncharacteristic concision lies in the fact, as Röhl himself points out, that the Kaiser's significance "*drastically declined*" with the outbreak of war:

---

15 *W. Görlitz*, The Kaiser and His Court. The Diaries, Letters and Notebooks of Admiral Georg Alexander von Müller Chief of the Naval Cabinet 1914–1918, New York 1959, 26, 28, 40, 41.

16 *J. C. G. Röhl*, Wilhelm II. Der Weg in den Abgrund, 1900–1941, Munich 2008, 1186.

17 Ibid., p. 1186.

18 *Afflerbach* (note 7), 44–5, 53, 83.

19 *Cecil* (note 11), vol. 2, 219.

> *"Power passed with astonishing speed from the Kaiser to the supreme military command."*[20]

It would be a mistake, however, to push this argument too far. The nervousness and lability of the sovereign, the efforts of the most senior officials to protect him from unwelcome news, the whole repertoire of tricks and devices that Geoff Eley once summarised with the phrase "*managing the Kaiser*" – all this was nothing new. On the other hand, the imperial *office* as a constitutional fact remained unchanged after the outbreak of war. Wilhelm could and had to be crowded out of the operative sphere of command over the land forces – in this area, the ambition of the universalizing monarch had to yield before the custodians of expert knowledge. But he still occupied a position hinge at the constitutional hinge between the military and the civilian authorities. Indeed, he represented the sole agency in which the military and the civilian chains of command merged.

Above all, the Kaiser retained the power to appoint and dismiss 'his' officers and officials. This was a matter of little practical import for the great mass of military personnel decisions, but the appointment to Chief of the General Staff, the highest office of wartime operational command, was another matter. Wilhelm had traditionally seen this post as an issue for his own personal decision and did not necessarily feel bound by considerations of seniority or by the recommendations of the Military Cabinet. He had been closely involved in the appointment of Helmuth von Moltke in 1906, a decision that surprised many informed observers because Moltke, though of senior rank, had never served as chief of staff at army corps level[21]. But clashes over policy during the July Crisis had strained relations between the two men and in September 1914, when Moltke suffered a nervous breakdown after reports arrived of serious setbacks for the German forces at the front, Wilhelm played a crucial role in his dismissal[22]. Wilhelm's Adjutant-General and Commander of the Imperial Headquarters, Colonel General Hans Georg von Plessen, described in his diary the dramatic final scenes of Moltke's career as Staff Chief:

> *"Moltke is summoned together with Falkenhayn. His Majesty says to Moltke that in view of his fragile condition he should report sick, the state of his nerves called for a break – 14 days! Moltke says: 'I won't do it! I am not sick. If Your Majesty is not satisfied* [with me] *then I will go!"*[23]

---

20 *Röhl* (note 16), 28.

21 *Annika Mombauer*, Helmuth von Moltke and the Origins of the First World War, Cambridge 2001, 45.

22 *Afflerbach* (note 12), 187.

23 Cited in *J. C. G. Röhl* (note 16), 1182–3.

His Majesty was *not* satisfied. On the following day he issued a Cabinet Order transferring the tasks and responsibility of CGS to Erich von Falkenhayn. In choosing Falkenhayn, for whom he had long shown a special personal regard, Wilhelm once again flouted the preferences of many senior military figures.

Falkenhayn was a highly controversial appointment and he was to become increasingly unpopular as the German forces on the western front failed to break the Allied lines in the winter of 1914. The very fact that Falkenhayn's continuing tenure in office was contingent upon Wilhelm's personal support accentuated the element of dependency in the relationship. In this sense, as Falkenhayn's biographer Holgar Afflerbach has pointed out, the Chief of the General Staff remained a "*court favourite*". The new CGS was an uncommonly gifted communicator – an important attribute in Wilhelm's eyes – whose Vorträge stood out for their brilliance and lucidity. He also possessed the quality known in military circles as 'Schneid' – the ability cut a sharp, soldierly figure. The Kaiser's tenacity in retaining this embattled commander had more to do with a reluctance to let go of a familiar and trusted subordinate than with an ambition to regain operative control. Nonetheless: this was an intervention that underscored the emperor's continuing centrality within the power structure[24].

By the end of 1914, there was growing pressure from within the military to dismiss Falkenhayn. The root of the anti-Falkenhayn agitation – aside from professional jealousies and rivalry – lay in the polarization of views on how the war should be conducted. Falkenhayn believed that the key to ultimate success lay in the west, and that it was here, where Germany faced the combined forces of France and Great Britain, that she must commit the bulk of her resources. In a long memorandum of December 1915, he explained that if the German army were to select and focus on a target that the French would feel compelled to defend until the last, it would be possible, by dint of sheer attrition, to "*bleed white*" the French army. The fruit of this strategy was the colossal assault launched by the German army in February 1916 against the fortifications around the fortress of Verdun. By contrast, Hindenburg and Ludendorff saw the key to a German victory in the complete destruction of the Russian forces in the east. They felt themselves vindicated by the relatively successful record of the German armies on the Russian front and complained that they were being starved of resources in order to sustain a westward policy that had not yet delivered a decisive

---

24 *K.-H. Janssen*, Der Kanzler und der General. Die Führungskrise um Bethmann-Hollweg und Falkenhayn (1914–1916), Göttingen 1967, 58, 62; *Afflerbach* (note 12), 237, 238–40.

breakthrough. In the summer of 1915, there was an open dispute over these questions between the eastern commanders and the CGS[25].

As the campaign to force the general out of office gained momentum, Wilhelm became increasingly aware that his own authority, as well as Falkenhayn's, was at stake. The threat to Wilhelm's independence was made clear when the dominant figure in the anti-Falkenhayn agitation, Field Marshal Paul von Hindenburg, threatened to resign unless Falkenhayn were dismissed. Hindenburg and his close associate Erich Ludendorff, a nominal subordinate but the more gifted strategist and organizer of the two, were the dominant commanders and strategic planners of the Eastern Front, and were widely credited with the spectacular victory over the Russian armies at Tannenberg, East Prussia, in August 1914. Hindenburg's ultimatum set a new precedent in the history of the Prussian military – never before had an officer attempted to trade his continuation in service for the pursuit of a particular policy. Wilhelm was furious at this manipulative behaviour and considered bringing Hindenburg before a court martial. He refused the resignation and failed to discipline the unruly Field Marshal, presumably because he feared alienating such a prestigious and popular commander[26].

A deadlock of this kind could not be resolved without imperial arbitration. The situation was made more difficult for Wilhelm by the fact that the men of the Eastern Command and their allies within the central command structure were supported by no less a figure than the Imperial Chancellor Theobald von Bethmann Hollweg. Indeed, it was Bethmann who coordinated the early phases of the anti-Falkenhayn agitation. Bethmann rejected Falkenhayn's view that the key to success lay in the combination of a massive assault on the western front with diplomatic overtures to Russia (the responsibility for conducting which would lie with the Chancellor). He supported the Hindenburg-Ludendorff view that Germany would only be able to secure a favourable outcome if the strategic fulcrum of the struggle were shifted from west to east. The anti-Falkenhayn camp even succeeded in recruiting members of the royal family, among them the Kaiser's youngest son Joachim, then serving in Hindenburg's headquarters, and the Empress herself, who agreed at Hindenburg's prompting to write to her husband in support of Falkenhayn's dismissal[27]. In a telegram to Auguste Viktoria, Wilhelm expressed his indignation at the fact

---

25 *Cecil* (note 11), vol. 2, 227–8; *Afflerbach* (note 12), 307–8.

26 *H. Herwig*, The First World War. Germany and Austria-Hungary 1914–1918, London 1997, 134; Görlitz (note 15), 57.

27 *Wolfram Pyta*, Hindenburg: Herrschaft zwischen Hohenzollern und Hitler, Munich 2007, 156, 163, 164.

> *"that forgers of intrigues have not stopped short of My House, but rather, with an impertinent disregard for its peace, have dared to send even you into the field against me"*[28].

In the dispute of July–August 1915, Wilhelm took a mediating position. Falkenhayn was ordered to move new troops from the western front to the eastern command, but he was retained in office over the protests of Hindenburg and Ludendorff, and Hindenburg was ordered to operate within the framework of Falkenhayn's strategic concept. Hindenburg's authority was cut back by the creation of a new civilian administration within the area of Russian Poland that had previously under the control of the Eastern Command. But this was merely a temporary respite for the embattled CGS. The gigantic offensive launched against the Verdun fortress system in February of the following year dragged on into the summer at a huge cost in German (and French) lives without achieving a breakthrough. On 30 July 1916, Hindenburg and Ludendorff successfully pressured Wilhelm into approving a concentration of control on the eastern front that substantially undercut the authority of Falkenhayn's supreme army command[29]. The death-knell for Falkenhayn came with Rumania's unexpected entry into the war on the Allied side on 28 August 1916. The emergence of a new and apparently formidable enemy in the east shifted the fulcrum of the war-effort towards the eastern front and cast serious doubt on the appropriateness of Falkenhayn's westward-oriented strategy. On 30 August 1916, one day after Rumania's entry into the war, Wilhelm bowed to the inevitable and appointed Hindenburg chief of the general staff, with Ludendorff as his quartermaster-general[30].

Falkenhayn had warned Wilhelm that the appointment of Hindenburg and Ludendorff to the supreme command would effectively nullify his sovereignty as emperor[31]. When Lyncker, as Chief of the Military Cabinet, with responsibility for personnel decisions, proposed that the command structure should be concentrated and placed in the hands of Hindenburg, Wilhelm protested that this would mean

> *"abdication for him [... It would mean that] Hindenburg had in effect usurped his place as a tribune of the people"*[32].

---

28 Wilhelm II to Auguste Viktoria, 21 January 1915, cited in *Pyta* (note 27), 164.

29 *Cecil* (note 11), vol. 2, 236; *Herwig* (note 26), 195–6; *Afflerbach* (note 12), 431–439; *Falkenhayn* (note 9), 272–3; *Afflerbach* (note 7), 49.

30 For details, see *Karl-Heinz Janßen*, Der Wechsel in der Obersten Heeresleitung 1916, in: Vierteljahrshefte für Zeitgeschichte 7 (1959), 337–71.

31 *Afflerbach* (note 12), 429; more generally, see *R. Pommerin*, Der Kaiser und Amerika. Die USA in der Politik der Reichsleitung 1890–1917, Boppard 1986, 334–76.

32 Admiral Müller, diary entry of 3 July 1916 in: *Georg Alexander von Müller*, Regierte der Kaiser? Kriegstagebücher, Aufzeichnungen und Briefe des Chefs des Marine-

It is certainly true that the two generals quickly ate into what little remained of Wilhelm's role as military commander-in-chief. A new minister of war, Hermann von Stein, was selected for his pliancy and political loyalty to the supreme command. The Deputy Commanding Generals, officers with extensive domestic administrative powers who had hitherto reported directly to the emperor, were placed under the authority of the minister of war. Nevertheless, it would still be mistaken to conclude that Wilhelm could no longer play a significant role in the executive structure as a whole. The promotion of the eastern generals had closed the gap in the military command structure by imposing unity of control from above, but there remained important divergences between the civilian and military leadership that periodically required the emperor's adjudication. The most important of these during the middle war years concerned the deployment of submarines against enemy shipping.

For the German political and military leadership, the use of submarines was the most contentious issue of the war. The naval command argued that the submarine could play a decisive role within the German war strategy by establishing a counter-blockade against Britain and thereby forcing her out of the war. Achieving this would entail deploying submarines against merchant vessels of the enemy belligerents and warning neutral merchant craft away from the 'war zone'. But the deployment of submarines in this fashion was fraught with risks. International law did not provide for the use of submarines against mercantile craft. The law provided for the interception of enemy surface warships that were supposed to follow the rule of 'visit and search'. But if neutral craft (including those of neutral states) continued to traverse the war zone, were they fair game? Even if they were declared exempt from attack, accidents were inevitable, since submarine captains often found it difficult to distinguish between neutral and enemy vessels. There was also the problem of defining and distinguishing legitimate targets. Did an exemption granted to passenger vessels still apply if these were carrying contraband military supplies? Did Britain's 'illegal' blockade against Germany's northern ports justify reprisals against British non-military shipping? And what was the status of those craft that sailed under a belligerent flag but also carried passengers of neutral nationality?

By the beginning of 1915, the question of how U-boats should be deployed had sown deep divisions within the decision-making elite. On the one hand there were those, like Bethmann, the Cabinet Chiefs Müller and Valentini, and Treutler, the foreign affairs liaison official at Imperial Head-

Kabinetts Admiral Georg Alexander von Müller 1914–1918, ed. By Walter Görlitz, Göttingen 1959, 200.

quarters, who argued that the risks of a comprehensive U-boat campaign outweighed the prospective gains. The crux of their argument was that if indiscriminate U-boat attacks were to bring America into the war on the side of Germany's enemies, Germany's prospects of victory or even an advantageous peace would evaporate. Opposed to this group were the naval and military hawks around state secretary Tirpitz and the naval chief of staff Bachmann, who saw maximum submarine deployment as the only means of retaliating against the British 'hunger blockade'[33].

Wilhelm was ambivalent on the matter of submarine deployments. He initially favoured the moderate view, largely because he feared the consequences of alienating the United States, but also because, as he later put it, "*the idea of innocent passengers drowning at sea appalled him*"[34]. In February 1915, however, after learning that the Americans were supplying the Allies with submarines and other military equipment, he agreed under pressure from Tirpitz that Allied merchant vessels within a designated war zone could be torpedoed without warning. Launched on 4 February, the new policy yielded disastrous results in early May when the *Lusitania*, a large passenger liner with contraband cargo in its hold, was torpedoed and sunk. Among the 1,198 passengers drowned were more than one hundred American citizens. Unnerved by the international uproar that followed and desperate to prevent an American entry into the war, Bethmann and Müller pressed for the re-imposition of restrictions. After a meeting of the crown council on 31 May 1915, Bethmann prevailed over the submarine enthusiasts, and on the following day (1 June 1915), Wilhelm issued an order to all submarine captains that neutral shipping was henceforth to be spared and that, in cases of doubt, "*it is better to let an enemy merchantman pass than to sink a neutral*". A further order issued on 6 June 1915 forbade all attacks on large passenger ships of whatever nationality[35].

Tirpitz and Bachmann were furious and responded with a joint telegram setting out their objections to the new policy, but Wilhelm stuck to his ground and added that he expected the order to remain absolutely secret – a precaution intended to prevent the demagogical Naval Secretary from agitating publicly against the government's restrictions. In an ominous re-run of Hindenburg's insubordination of January, Tirpitz and Bachmann now submitted their resignations. Wilhelm refused to grant them, exclaiming to an adviser:

---

33 Pommerin (note 31), 334–76.

34 Conversation with Holtzendorff reported in Müller diary, 15 January 1916, in *Görlitz* (note 15), 126; on William's motivations, see also *Cecil* (note 11), vol. 2, 221.

35 Cited in *E. R. May*, The World War and American Isolation 1914–1917, Cambridge Mass. 1959, 208

> *"No! The gentlemen have to obey and to remain. A regular military conspiracy! Brought about by Tirpitz."*[36]

After the sinking of a further passenger ship, the *Arabic*, mistaken for a freighter by Lieutenant Schneider of U-24 on 19 August 1915, Bethmann persuaded Wilhelm to introduce further limitations on deployment, thereby rendering the submarine weapon "*virtually impotent*"[37]. Tirpitz was predictably outraged and once more submitted his resignation. Wilhelm was reluctant – as earlier in the case of Hindenburg – to dismiss an official who enjoyed such widespread public support, but he did dismiss Tirpitz's irritating sidekick, Bachmann. In a letter to the Naval Secretary that drew on the arguments of the political advisers grouped around the chancellor, Wilhelm explained that preventing America from entering the war as an

> *"active enemy" was of supreme importance, since "she could provide unlimited money for our foes". "As Chief Warlord I had absolutely to prevent this* [...] *First the war must be won, and that end necessitates absolute protection against a new enemy; how that is to be achieved* [...] *is My business. What I do with My navy is My business only."*[38]

Throughout 1916, despite increasingly widespread parliamentary and press agitation for unrestricted submarine warfare (USW), Wilhelm and the Bethmann group kept the submarine lobby at bay. On 15 March, Bethmann persuaded Wilhelm to strip the naval secretary of some of his ministerial responsibilities. Tirpitz's resignation – the third since the outbreak of war – was now accepted by the Kaiser. After a brief experimental period of slightly loosened restrictions produced yet another controversial sinking (the cross-channel steamer *Sussex*, mistaken for a mine-layer on 24 March), Wilhelm issued a new order stating that no vessels whatsoever (excluding warships of the belligerent powers) were to be sunk without prior warning, which led the naval commanders to suspend submarine operations entirely in the Atlantic and English Channel. The Kaiser thus – at least for the moment – fulfilled the decision-potential of his office in holding the fort against the advocates of USW. His position was motivated in part by geo-strategic concerns – especially as regarded the incalculable consequences of a US entry into the war – but strong moral reservations were also in play. He took the view

36 Cited in *May* (note 35), 209.

37 *May* (note 25), 222.

38 Cited in *May* (note 35), 223; the economic arguments summarised here are those of the Treasury minister Karl Helfferich.

> *"that to torpedo great ships full of women and children is a barbaric cruelty without parallel that would arouse the hatred and the venomous anger of the entire world against us"*[39].

Wilhelm thus played a crucial role in supporting the politically moderate elements within the decision-making elite against the demands of the hawks. Why, then, did he endorse a policy of unrestricted submarine warfare in January 1917? Public opinion was certainly a key factor. During the second half of 1916, there was increasingly vocal and widespread support for submarine warfare, especially after the 'Battle of Jutland' at the end of May failed to lift the British blockade. Support for USW now came not only from the annexationist milieu that had dominated the submarine agitation of 1915, but also from across the spectrum of the Reichstag parties. It appealed to a German public whose hostility was increasingly focussed upon Great Britain as the keystone of the enemy coalition[40]. The submarine became, as Roger Chickering has put it:

> *"a panacea, the wonder weapon whose all-out employment promised to resolve the war and bring the British to their knees"*[41].

The prevailing climate of opinion was not, of course, in itself sufficient to convert Wilhelm to a policy of USW, though its impact on him can be scarcely be doubted. But it weakened the position of the chancellor and thereby helped to tilt the balance within the executive towards the submarine enthusiasts. During the summer and autumn of 1917, enthusiasm for USW seeped like a powerful solvent through the chancellor's parliamentary base. The conservatives and the national liberals broke away to form a 'U-boat bloc'; by October 1917, even the formerly loyal Centre Party was openly opposing the chancellor's policy on submarine deployment and demanding that the power of decision be placed in the hands of Hindenburg and Ludendorff, whose support for USW was widely known. These developments undermined the chancellor's confidence and personal standing in the tough debates within the executive upon which everything depended.

A further important factor in Wilhelm's decision to accept USW was the failure of Bethmann's peace sounding to the nations fighting Germany in December 1916. The peace note of 12 December was the product of a fierce struggle over wording between the chancellor and the supreme command. The result was a characteristically ambivalent communiqué that opened with an announcement that Germany was willing to talk about peace, but closed

39 Hans Georg von Plessen, diary entry of 10 January 1916 in *Afflerbach* (note 7), 847.
40 See *M. Stibbe*, German Anglophobia and the Great War, 1914–1918, Cambridge 2001.
41 *Chickering* (note 11), 92.

with a warning that she would fight onward to victory if discussions did not take place. Wilhelm gave his full support to Bethmann's initiative, not merely for reasons of 'expediency' as Lamar Cecil has suggested[42], but also because he was genuinely weary of the war (he is reported to have burst into tears at the mention of the word 'peace' during a meeting with a Reichstag delegation[43]) and because the making of peace also accorded with his sacralized conception of the sovereign's office: peacemaking, he told Bethmann, was a

> *"moral act that was appropriate for a monarch who has a soul, and feels himself responsible to God, who has a feeling for his people and the enemy's"*[44].

In the event, the peace initiative of 12 December was a failure. Five days later, Admiral Müller observed that Wilhelm's buoyancy at the news of German victories against Rumania had given way to gloom,

> *"chiefly because foreign statesmen have looked down their noses at our peace feelers"*[45].

When the Allies' negative reply to the peace note was published, Wilhelm grew angry, declaring that Germany must fight on to the end, annex Belgium, subdue France and so on.

In the fatalistic, self-righteous mood of late December 1916, Wilhelm, whose commitment to the Bethmann view had in any case always been ambivalent and conditional, was disposed to embrace the promise and risks of unrestricted submarine warfare. The submarine enthusiasts, for their part, had raised their act since 1915; the stock of sumbmarines had been enlarged and improved and the enthusiasts could now produce barrages of tables and statistics demonstrating that five months of ruthless submarine warfare against Great Britain's commerce in the Channel and the Atlantic (submarine warfare had never been suspended in the North Sea) would suffice to drive her out of the war. The risk of an American entry into the war was irrelevant, they contended, since German submarines would already have decimated Allied transatlantic traffic before substantial numbers of US troops were ready for embarkation and would in any case have little difficulty in stopping troop shipments from the United States to Europe. Wilhelm quickly adopted these arguments as his own. At a meeting of 9 January 1917, organized by the military command to discuss the issue in the eastern

42 *Cecil* (note 11), vol. 2, 241.
43 *May* (note 35), 398.
44 Wilhelm II to Bethmann, 31 October 1916, cited in *May* (note 35), 398–9; *Cecil* (note 11), vol. 2, 242.
45 Müller diary, 17 December 1916, in *Görlitz*, (note 15), 224.

front headquarters at Pless, it was apparent that the Kaiser had already resolved to support the generals. Bethmann had come armed with arguments against a new submarine campaign, but found himself isolated and eventually acquiesced in the majority view. Wilhelm showed signs of impatience as Bethmann enumerated his objections, and then signed an order to the effect that unrestricted submarine warfare was to commence from 1 February. When this decision was formally announced by the German government, it resulted in the almost immediate severance of diplomatic relations with the United States and in April in the latter's declaration of war on Germany.

At the meeting in Pless, Wilhelm had presided over a decision of world-historical moment. It was founded, as historians have pointed out, on an egregious miscalculation of the risks and benefits involved. Although German submarines reached, and for a time even exceeded, the projected target of sinking 600,000 tons a month, the advantage to Germany was shortlived. The British food supply and distribution system proved more elastic than expected, the capacity of the American shipyards to replace lost tonnage was much greater than the German strategists had appreciated, and the anti-submarine measures developed by the Allies proved more effective than the Germans could have predicted. The loss rate among German submarines soared, and soon exceeded that among Allied convoys by a factor of seven to one in percentage terms[46].

Of course, it could be argued that Germany had no other choice, because in the long run the odds in a continental war of attrition were stacked against Germany. But was this true? It is open to doubt, to say the least. In January 1917, Germany had just succeeded in crushing Rumania and victory in Russia was not far off, though the Germans could not know that. The morale of the French army was close to collapse and Britain was fast running out of money, indeed it was much closer to financial collapse than the Germans knew. During the autumn of 1916, American exasperation at the British blockade against Germany was rising and Anglo-American relations were at their nadir. Without American participation and the comprehensive aid that came with it, Britain may well have had to sue for peace in the summer or autumn of 1917, at around the time that the Italian front was beginning to collapse under Austro-German pressure.

In other words: if Germany had not embarked on unrestricted submarine warfare against merchant shipping and the United States had stayed out of the war, a German defeat at Allied hands seems highly unlikely. Germany's best chance – seen in retrospect – lay in

46 *Ferguson*, The Pity of War, London 1998, 283; *Chickering* (note 11), 93; *Pommerin* (note 31), 364.

> "*simply waiting for the paralysis of* [Allied] *shipping, finances and military collapse on several fronts*"[47].

Wilhelm himself appears temporarily to have glimpsed this possibility. During a visit to Vienna late in November 1916, he predicted "*revolution in Moscow and St Petersburg*", "*complete munitions failure in Russia*", "*famine in England*", and a French army worn down "*to the last man*"[48]. Wilhelm's interlocutors greeted this outburst with weary scepticism, but as a vision of a 'virtual future', it was less implausible than they thought.

During the autumn and winter of 1916/17, the Kaiser's power was further eroded by a sequence of political and military crises. There was a drastic extension of the power wielded by the military over the civilian arm of government. In a series of clashes with the generals, Bethmann was forced to give way on important questions pertaining to the future status of Russian Poland and to accept the high-handed dismissal of various ministers and senior aides (Jagow, Helfferich, Hammann). By the beginning of 1917, notwithstanding Bethmann's role in helping to bring down Falkenhayn and hoist Hindenburg into the supreme command and his ultimate compliance over the submarine question, Hindenburg and Ludendorff had decided that the chancellor himself would have to go.

At the same time as these developments were unfolding within the executive, an increasingly volatile domestic political situation threatened to overwhelm the government from below. The outbreak of war had been followed by a brief period of 'national unity', in which the Reichstag factions agreed to exercise self-restraint on issues that divided the population. But by the summer of 1915, this 'truce' (Burgfrieden) had come under pressure from both ends of the political spectrum. On the left, a radical faction within the Social Democratic Party broke party discipline to denounce the war and vote in the Reichstag against further war credits. During 1915, there were increasingly frequent calls from the SPD for social and political rewards for the working class that was sustaining the fighting in the trenches. Key themes in left-wing political agitation were the rejection of a 'war of conquest' and the call for franchise reform. On the right, there emerged an influential ultranationalist network with powerful patrons in army, navy and government,

---

47 On the financial exhaustion of the Allies, see *J. M. Cooper Jr.*, The Command of Gold Reversed: American Loans to Britain, 1915–1917, in: Pacific Historical Review, 1975, 209–30; here 228; a similar argument is made by *G. Ritter*, The Sword and the Sceptre: The Problem of Militarism in Germany, trans. H. Norden, Coral Gables 1972, vol. 3, 325.

48 Report from Wedel, the German Ambassador in Vienna, cited in *W. Gutsche*, Wilhelm II der letzte Kaiser des Deutschen Reiches : eine Biographie, Berlin 1991, 181.

which demanded extensive German annexations as the sine qua non of peace and opposed domestic political reform.

During 1916, the right had briefly succeeded in rallying a substantial body of parliamentary support around the campaign for the submarine. But by the early spring of 1917, the political initiative had passed to the left. Severe food shortages and news of the February Revolution in Russia generated a highly volatile mood in the German industrial cities and the centre-left was beginning to coalesce around the call for domestic political reform, and specifically for the abolition of Prussia's antiquated and discriminatory three-class franchise. In order to restore calm, Bethmann urged Wilhelm to issue an 'Easter Message' promising that franchise reform would follow the cessation of hostilities. Wilhelm had – albeit reluctantly – accepted the need for democratization since early in 1915 and was willing to comply[49]. But his vaguely-worded promise merely raised the political temperature. The following weeks saw the first major strikes since the beginning of the war, and renewed efforts by the Reichstag to assert political control over the German war effort. Many deputies now insisted upon franchise reforms without further delay, and much of the early summer was spent discussing a 'Peace Resolution' moved by the prominent Centre deputy Matthias Erzberger, which called for a negotiated peace without forced annexations.

Wilhelm and Bethmann found themselves in open conflict with the military leadership over both these issues, although it was the franchise question that supplied the occasion for the chancellor's fall from office. Early in July 1917, when Hindenburg and Ludendorff learned that Bethmann was pressing Wilhelm to announce immediate franchise reforms, the two commanders travelled to Berlin to demand the chancellor's dismissal. But Wilhelm held firm and Bethmann remained in office. Bethmann was now under pressure not only from the military command, but from the Reichstag deputies, a majority of whose deputies had lost patience with the chancellor's slowness to instigate franchise reform and were demanding his dismissal. The wily Ludendorff even attempted to discredit Bethmann by hinting to a number of key Reichstag fraction leaders that he himself had no objection to franchise reform and that it was the chancellor who was the chief obstacle[50]. On 10 July, Bethmann at last succeeded in winning the emperor around to his standpoint and two days later it was announced that the next Prussian Landtag would be elected under a new democratic franchise. On the following day, in yet another of the spasms of insubordination to which the German

49 *Cecil* (note 11), vol. 2, 246.
50 *Ullrich*, Die nervöse Großmacht: Aufstieg und Untergang des deutschen Kaiserreichs 1871–1918, Frankfurt 1997, 522–9.

miltiary command had become so prone, Hindenburg and Ludendorff telephoned their resignations to Berlin, explaining that they could no longer work with the chancellor. Wilhelm was furious at being blackmailed in the way and observed to the chief of his Naval Cabinet that this behaviour on the part of Prussian generals was the most unheard-of thing that had ever happened in the history of Prussia'. He assured Bethmann that he would not give way, but he felt himself to be in an "*impossible position*" – with their letters of resignation, the two commanders had "*put a gun to his head*"[51]. The chancellor, too, was in a precarious position – implacably opposed by the military leadership and lacking any secure platform in the Reichstag, he now depended solely upon the support of the Kaiser, whose attachment was beginning to flag under pressure. Worn down by his struggle with the titans and anxious to save Wilhelm any further agonizing, Bethmann resigned.

The resignation of 13 July marked a fundamental caesura in the history of the reign. Wilhelm saw this clearly enough: "*Now I may as well abdicate*", he told Bethmann[52]. Nothing better illustrates the dramatic collapse of what remained of Wilhelm's authority than the identity of the new chancellor. The first five chancellors of Wilhelm's reign had all been personally known to the Emperor before they served under him: Hohenlohe was his 'uncle', Bülow was a member of the Eulenburg circle, Wilhelm had shot his first stag in the company of Bethmann's father. By contrast, the new incumbent, Georg Michaelis, was a virtual stranger to Wilhelm, though not to his wife, who knew him through her church connections and was willing to speak up for him on that score. Michaelis was known as an efficient administrator – he had been responsible for organizing the civilian and military food supply – but he was hardly a prominent public figure, and he was not Wilhelm's choice for the job. Nor was Count Hertling, who replaced Michaelis at the Chancellery when the latter fell from favour with the Reichstag in October 1917. The Chancellorship – that pivotal office in the German constitutional system – was no longer in the emperor's gift. Wilhelm was still, on occasion, capable of standing up to the 'terrible twins', as in January 1918, when he opposed Ludendorff's plans to annex Poland to the Reich and headed off a subsequent resignation threat[53]. But the twins soon extracted their revenge by forcing the dismissal of trusted officials from within Wilhelm's personal advisory group, among them the faithful Cabinet Chiefs Valentini and Lyncker.

51 *Müller* (note 32), 304, 303 (entries of 16 and 13 July, 1917).
52 *Fehrenbach* (note 3), 218.
53 *Michael Balfour*, The Kaiser and his times, London 1964, 385–6.

We need not examine in detail the endgame of Wilhelm II's reign – a phase of opportunist, last-minute constitutional adjustments in the face of certain defeat. There was one last, desperate flexing of the monarch's constitutional muscles. In October 1918, as it became clear that the Allied peace conditions were going to be tougher than the German leadership had expected, Hindenburg reneged on his earlier agreement to delegate responsibility for negotiating peace to the civilian government and circulated a note to his generals declaring that Wilson's terms were unacceptable. Rightly perceiving this as a challenge to his own authority, the newly-appointed chancellor Max von Baden threatened to resign unless Wilhelm put an end to civil-military 'double government' in Germany. Wilhelm was thus obliged to deploy his sovereign prerogative in a final act of adjudication between the two power-centres. On 26 October, he upbraided Ludendorff and accepted his resignation.

> "*The operation is over*", he later remarked. "*I have separated the Siamese twins*"[54].

It was too late, of course, to save the Prussian-German throne.

*

In this chapter, we explored the tensions and asymmetry between the different kinds of power wielded by the last German Kaiser in wartime. The outbreak of war revealed the emperor's total inability to participate in any way in the operative leadership of the German forces. On the other hand, a significant residue of power remained in his hands on account of the central position he occupied in the complex joistwork of the Prussian-German composite monarchy. This constitutional dimension of his power, a default feature that remained intact even when the Kaiser ceded the political initiative to his most powerful subjects, has rarely been accorded sufficient weight. Of course, Wilhelm II was never able to form the power residue of his office into the basis for a genuine leadership role. He could have used his sovereign position in order to mediate between the inner fronts of the empire: between the army and the navy, between the civilians and the military leaders, between the central and the peripheral military commands; he might thus have been able to compensate for the absence in Germany of a unified structure of agency of command comparable with the Conseil Supérieure de la Guerre in France and the Committee of Imperial Defence in Britain. But for this Wilhelm II lacked the confidence, the consistency of intention and the breadth of vision. On the other hand, in several questions of

---

54 Ibid., 400.

fundamental importance, Wilhelm demonstrated a degree of determination and proved capable – at least temporarily – of impeding a disastrous policy initiative; on occasion he even displayed remarkably sound judgement. We should be wary of drawing too sharp a dividing line between the pre-war and the wartime Wilhelm II. The Kaiser was neither an all-powerful titan nor an inconsequential shadow-emperor – and that is as true for the years after the outbreak of war as it is for the years before.

Klaus Larres

Churchill: Flawed War Leader or Charismatic Visionary?

Winston S. Churchill was a controversial politician throughout his lifetime and has remained so ever since. The scholarly and more popular tomes written about him as leader in war and peace range from sheer admiration and hagiography to utter condemnation of his political and indeed military and strategic decisions. Not surprisingly, some authors depict him as a deeply flawed war strategist while others admire him as a highly charismatic and visionary leader. Both assessments are not necessarily contradictions in terms. This article will demonstrate that during the Second World War, as indeed in his subsequent career as peace time Prime Minister during the first half of the 1950s, Churchill can be aptly characterized as both a flawed war leader *and* a charismatic and persuasive leader. He possessed genuine visionary insights, at least on certain occasions. Following a brief perusal of the literature, Churchill's performance as war leader during World War II is assessed in this contribution.

Churchill once quipped that history would judge him gently as he intended to write that history himself. Yet, this well-known statement contained a large grain of truth. Until the early 1970s it was Churchill six-volume history of the Second World War, mostly written in the second half of the 1940s, which shaped the scholarly and popular writings on the course of the war to a considerable degree. Enriched by a plethora of quotations from a large number of primary documents, the war-time Prime Minister's authoritative outline of the development of the Second World War remained the most detailed and persuasive account for the better part of two decades. His elegant and persuasive literary style also earned him a Nobel prize for literature in 1953. Only with the gradual opening of western archives became it possible to challenge Churchill's self-serving argumentation, complement his account and indeed offer alternative historical narratives for some of the war's crucial turning points[1].

All major political and historical figures, and not least important war leaders, have their performance frequently revised. The reputation of very few of them is raised in the process. Churchill, however, has a solid circle of supporters, such as historian Martin Gilbert and others organized in the various Churchill societies round the world, whose judgment will always remain highly positive[2]. Among the latest offerings is Paul Johnson's populist and somewhat imbalanced account[3]. However, at the hand of more scholarly historians and indeed military experts Churchill has suffered and his reputation is increasingly looked at in a more critical light.

As Elliot Cohen has pointed out in an insightful article[4], the first wave of criticism came at a fairly early stage. Initially there were Britain's former military leaders who in their memoirs began criticizing Churchill's leadership as war strategist. For instance, Churchill's chief of the Imperial General Staff, Lord Allanbrooke, criticized Churchill severely in his diaries, parts of which were published in the late 1950s. He described Churchill as a war leader who did not sufficiently listen to his military advisers. Allanbrooke believed that essentially Churchill was a dangerous amateur who was too impressed by his own strategic insights[5].

The second major wave of criticism was driven by historians who had gained access to the primary western documents once they had become available in the course of the 1970s under the time-honored thirty-year-rule which largely still governs access to governmental records in Britain. In the accounts of these historians Churchill frequently was described as a "romantic militarist", someone who had "grand plans" and ideas but who often did not pay sufficient attention to the details and to the situations on the ground. He was also seen as meddling and bullying, and as a leader who had too much imagination and too many fanciful ideas. Many authors viewed him as only a semi-successful war strategist at best as he did not devise any coherent war strategy and was often carried away by his enthusiasm and idiosyncratic ideas[6].

Much of this and subsequent criticism was quite justified and made on the basis of deep scholarly research. Still, over the years there also occurred

1 *D. Reynolds*, In Command of History: Churchill fighting and writing the Second World War, London 2004.

2 *M. Gilbert*, Winston S. Churchill, Vols 6 and 7, London 1983.

3 *P. Johnson*, Churchill, New York 2009.

4 *E. Cohen*, Churchill asks a question, in: *Supreme Command: Soldiers, Statesmen and Leadership in Wartime*, ed. by E. Cohen, New York 2002, 95–132.

5 *Lord Allanbrooke*, Allanbrooke War Diaries, new ed., London 2002.

6 For an overview, see *Cohen*, Churchill asks a question, 98 ff.

a great deal of unjustified attacks on Churchill's war leadership which made their way into books and articles. Not least John Charmley's account in the early 1990s comes to mind. Charmley argued somewhat unrealistically that Churchill should have arrived at a deal with Hitler in the course of 1940; that way he would have preserved the British empire instead of wasting Britain's resources by fighting World War II[7]. U.S. politician Pat Buchanan's 2008 book on Churchill argues along very similar lines. Buchanan also employs a number of quotations by Churchill and others to prove his points though he frequently distorts the meaning of these statements by quoting them only partially or out of context[8].

One of the most recent scholarly critics is Christopher Catherwood's book, published in 2009, *Churchill: The Flawed Genius of World War II.* Catherwood argues mainly that Churchill was wrong in delaying the invasion of Europe from spring 1943 to June 1944, an argument that has been often made before but still lacks conviction[9]. Recent much superior books that view Churchill's war-time performance in a more constructive way include Carlo D'Este's military biography of Churchill "Warlord" and Douglas Porch's "Path to Victory", an impressive overall history of the Second World War in the Mediterranean theatre of war[10].

Among some of the major, but often overlooked factors that influenced Churchill's decision-making during World War II are the following.

Churchill's personal memory of the First World War: the slaughter on the western front and the apparent limitations of Britain's senior military leaders had made a deep impact on Churchill's strategic thinking. His personal memory of World War I clearly influenced his thinking and decision-making during the Second World War.

The workings of governmental and military bureaucracies: Churchill distrusted the cumbersome and often inefficient bureaucracies, including Britain's military leadership and the Foreign Office. He would have preferred to do it all by himself rather than delegate important issues to some innocuous civil servants.

Deep mistrust of long-term planning: During the war Churchill paid great attention to detail with regard to the fighting of World War II but he always remained deeply distrustful of long-term planning, including strategic war-planning. With hindsight, Churchill's prolonged and unfortunate neglect of

---

7 *J. Charmley*, Churchill the end of glory: a political biography, London 1992.

8 *Pat Buchanan*, Churchill, Hitler and the Unnecessary War: How Britain lost its Empire and the West lost the World, New York 2008.

9 *Ch. Catherwood*, The Flawed Genius of World War II, New York 2009.

10 *C. D'Este*, Warlord: A Life of Churchill at War, 1874–1945, London 2009; *D. Porch*, The Path to Victory: the Mediterranean Theater in World War II, New York 2004.

planning for the post-war era was perhaps even more important. In London serious post-war planning was embarked upon considerably later than in Washington and Moscow.

But perhaps the following two factors were of particular importance with regard to how Churchill approached his role as war leader.

1. Churchill's Long Political Experience

Churchill had first entered parliament in 1900 and with only a brief interruption between 1922 and 1924, he remained a member of the British parliament until 1964. Before the outbreak of the First World War Churchill served several conservative and liberal Prime Ministers as Under Secretary of State for the Colonies, President of the Board of Trade, Home Secretary and First Lord of the Admiralty; he resigned the latter post as a result of the ill-fated Gallipoli campaign. Towards the end of World War I Churchill was recalled to government as Minister for Munitions and subsequently he served as Secretary of State for War and Air, Colonial Secretary and Chancellor of the Exchequer. His wilderness years, when he was out of office, lasted from June 1929 to 3 September 1939 when upon Britain's declaration of war on Nazi Germany Prime Minister Neville Chamberlain asked him to join the cabinet again as First Lord of the Admiralty. On 10 May 1940 Churchill succeeded Chamberlain as Prime Minister of a national coalition government.

Based on his long political experiences and his outstanding intellectual talents, Churchill arrived at the conviction that he was a superior military strategist than many of the military professionals whose task it was to design the details of Britain's war strategy. It is clear, however, that Churchill committed many major military and strategic blunders and the scholarly literature has rightly taken him to task for this for several decades now. For instance, his determination to embark on amphibious operations led to major catastrophes such as in Norway, West Africa, in Greece, Crete and Rhodes.

Still, Churchill also was responsible for some of the war's major strategic breakthroughs. Often he got it right. For instance, the Prime Minister was absolutely correct with his insistence on first allowing the new American and British armies some time to develop their professionalism and gain experience before sending them into a decisive battle. They urgently needed both to be able to take on the German armies in western Europe by means of the Normandy invasion. Churchill's strategic vision for Britain's role in the Battle of the Atlantic, including the crucial climax in the convoy battles in May 1943, was also of vital importance. It made it possible to get the millions of US ground troops into the UK and then further into Europe to liberate the continent. Thus, on the whole, Churchill's insistence on delaying the Second Front and thus the Normandy invasion was based on superior reasoning. Churchill was also prepared to resist strong American pressure and Stalin's fury until 1944. If there had been a Normandy landing in 1943 this only could have happened on a much smaller scale. Moreover, the strength of the German armies would not yet have been worn down in the huge battles against the Red Army that took place on the eastern front in 1943. Still, Churchill was also in favour of a delay of the Normandy invasion for quite selfish reasons in order to maintain the Mediterranean area open to British influence for example.

Thus, despite some major blunders, Churchill managed to successfully implement a great number of crucial strategic insights that were vital for allied warfare. He was frequently not deterred by strong opposition from his own chiefs of staff and, as in the case with the Second Front, from his American and Soviet allies.

Not least, in the trying, even desperate months of Britain "standing alone" in 1940–41, Churchill used his powerful rhetoric and his imaginative use of language for giving the British people and not least himself a great deal of hope and inspiration. He touched the British people's heart and also managed to greatly "*move the emotions*" of the American people. By appealing to liberty, a common humanitarian instinct and a nobility of instinct he "*also set the moral climate of the war*" as John Keegan has argued[11]. This may well have been one of Churchill's most decisive contributions to the survival of his country. As American journalist Ed Murrow wrote: "*Churchill mobilized the English language and sent it into battle*" to proclaim that the British people would never ever surrender and not allow themselves to be worn down by their enemies[12].

2. Cooperation with Allies

Perhaps one of Churchill's greatest strategic insights was the realization that Britain needed an alliance with above all the United States and then also with the Soviet Union to have a chance of survival. Churchill's conviction that he possessed unparalleled personal powers of persuasion and charisma came in

11 *John Keegan*, Churchill, London 2002, 134.

12 See for example *Ralph Keyes*, Ask not where this quote came from, Washington Post, 4 June 2006: http://www.washingtonpost.com/wp-dyn/content/article/2006/06/02/AR2006060201406.html

handy as he realized that he needed to do his utmost to keep the anti-Hitler coalition with the United States and the Soviet Union together[13].

By the time Churchill became Prime Minister on 10 May 1940 he had been active in British politics for forty years. Naturally he was fully aware of his unique personality and the persuasive impressions he was able to make on other people. We may wish to refer to this as Churchill's charisma. Due to his persuasive powers Churchill also very much believed in old-style diplomacy: personal negotiations with as few advisers present as possible. However, this did not make him conclude that he ought to negotiate with Hitler personally or enter into talks about a compromise peace with Hitler or representatives of the Nazi regime. The only time he attempted to talk to Hitler was in November 1932 in a hotel in Munich when Churchill was in the city as part of his research of a biography of his ancestor the Duke of Marlborough. Via an intermediary Hitler appeared to have actually agreed to meet the prominent British politicians, who, however, was out of office at the time. Hitler did not come to the meeting; it appears that he got cold feet[14].

During the Second World War Churchill always attempted to exploit fully his personal charisma in his relations with both American President Franklin D. Roosevelt and Soviet dictator Joseph Stalin. During as well as after the war Churchill's primary aim remained to maintain the great power position of the British Empire. His personal summit diplomacy was meant to facilitate the achievement of this goal while also allowing him personally to remain at the centre of international attention. In particular during 1940–41 his influence and powers of persuasion were decisive in securing Washington's assistance for a beleaguered Britain[15].

In September 1939, only eight days after the declaration of war on 3 September and nine days after Churchill had again been appointed First Lord of the Admiralty, President Roosevelt suggested to him that they engage in an occasional correspondence with one another. Roosevelt had realized Churchill's potential as future Prime Minister. Similar to Churchill Roosevelt believed deeply in personal diplomacy and his own powers of persuasion. Churchill accepted Roosevelt's offer immediately[16]. Churchill had

---

13 *Klaus Larres*, Churchill's Cold War: The Politics of Personal Diplomacy, New Haven & London 2002, chapter 2.

14 For this failed meeting, see *E. Hanfstaengl*, Hitler: The Missing Years, London 1957, 184; *P. Metcalfe*, 1933, Pb. ed., London 1990, 47–77; and for *Churchill*'s own account The Second World War Vol.1: The Gathering Storm, London 1948, 75–76.

15 For the following account of the development of Churchill's relations with both Roosevelt and Stalin, see also chapter 2 of my book Churchill's Cold War.

16 See *Churchill & Roosevelt: The Complete Correspondence, Vol. 1: Alliance Emerging*, ed. by W.F. Kimball, London 1984, 6–7, 23–25; also *J.P. Lash*, Roosevelt

spotted the opportunity to cultivate a personal contact important both for his country and his own career. He believed far too much in his political qualities to worry about the fact that he was virtually ignoring the Foreign Office as well as the British and American ambassadors in his high-level diplomacy with Roosevelt[17].

After Churchill's appointment as Prime Minister in a coalition government on 10 May 1940 and following the rapid and entirely unexpected German defeat of France and the entry of Italy into the war in June, the personal contact between the British and American heads of government became ever closer[18]. Despite the fact that the United States were still officially neutral, as early as September 1940 Roosevelt placed fifty destroyers at the disposal of the British government for the fight against the axis powers[19]. FDR's third consecutive election victory in November 1940 facilitated even more co-operation. The President was now able to pay less attention to the anti-war opinions that were expressed in many isolationist circles in the United States[20]. Moreover, due to the fierce resistance during the 'Battle of Britain' in August and October 1940[21], the British gained a tremendous degree of admiration and respect from the American public. Roosevelt ordered a gradual increase in the output of the American armaments industry. In January 1941 he presented the Lend-Lease Act to Congress. It would enable the British government, now desperately low on dollar reserves, to receive goods and weapons of all kinds from Washington without having

---

and Churchill, 1939–41: the Partnership that Saved the West, London 1976, 21 ff.; *D. Reynolds*, The Creation of the Anglo-American Alliance, 1937–41: a Study in Competitive Cooperation, Chapel Hill, N.C. 1981, 86–88.

17 See Kimball (ed.), ibid.

18 See *Reynolds* (note 16), 95 ff.; *Lash*, Roosevelt and Churchill, 93 ff., 154 ff. For good overviews regarding the Churchill-FDR relationship, see also for example *Barker*, ibid., 125–138; and *W. F. Kimball*, Wheel within a Wheel, 291–307; *W. F. Kimball*, Forged in War: Churchill, Roosevelt and the Second World War, London 1998; *J. Charmley*, Churchill's Grand Alliance: A Provocative Reassessment of the 'Special Relationship' between England and the U.S. from 1940 to 1957, New York 1995; and, with the emphasis on the intelligence dimension, *D. Stafford*, Roosevelt & Churchill: Men of Secrets, London 1999.

19 See *R. Hough*, Former Naval Person: Churchill and the Wars at Sea, London 1985, 157–62; *P. Goodhart*, Fifty Ships that Saved the World. The Foundation of the Anglo-American Alliance, Garden City, N.Y. 1965, 177 ff., and 3 ff. for the earlier history.

20 See *W.S. Cole*, Roosevelt and the Isolationists, 1932–45, Lincoln, Neb. 1983; also *W. Heinrichs*, Threshold of War: Franklin D. Roosevelt and American entry into World War II, Oxford 1988.

21 For a good interpretive account, see *R. Overy*, The Battle, Harmondsworth 2000; and for a detailed and exhaustive treatment, *S. Bungay*, The Most Dangerous Enemy: A History of the Battle of Britain, London 2000.

to worry, for the time being, as to how the shipments would be financed[22]. However, Britain became increasingly dependent on American goodwill. Also in January 1941 Roosevelt's confidant, Harry Hopkins, flew to London to make Churchill's acquaintance. Ambassador Averell Harriman as well as Under Secretary of State Sumner Welles paid a visit to Churchill in March. As these men had great influence over Roosevelt, Churchill was very keen to gain their trust and he did everything possible to establish a close relationship with all three[23].

Then, in August 1941 the first personal meeting between Churchill and Roosevelt took place aboard a ship in Placentia Bay near Argentia in Newfoundland. It led to the signing of the *Atlantic Charter*, a vague Anglo-American statement on common war aims which expressed the two leaders' "*hopes for a better future for the world*"[24]. At the same time, Anglo-American differences on self-determination, anti-colonialism, the future system of global trade and the post-war order already came out in the open to some extent. In the long run these disagreements would seriously affect the war alliance as well as Churchill's personal relations with Roosevelt[25].

---

22 See the still useful essay by *D. Waley*, Lend-Lease, in: *America, Britain, and Russia: their Cooperation and Conflict, 1941–46*, ed. by W. McNeill, London 1953, 772–89; and *W. F. Kimball*, 'The Most Unsordid Act': Lend-Lease, 1939–41, Baltimore 1969; also *Reynolds* (note 16), 145–68; *A. P. Dobson*, US Wartime Aid to Britain, 1940–46, London 1986; *W. F. Kimball*, The Juggler: Franklin Roosevelt as Wartime Statesman, Princeton 1991, 127–57.

23 See above all *R. E. Sherwood*, Roosevelt and Hopkins: An Intimate History, New York 1948, 176–77, 275 ff.; *W. A. Harriman* and *E. Abel*, Special Envoy to Churchill and Stalin, New York 1975, 56 ff.; *D. Stafford*, Churchill and Secret Service, London 1997, 197–99; for good accounts of the visits of Roosevelt's envoys, see also *M. Gilbert*, Winston S. Churchill, Vol. 6: Finest Hour, 1939–41, London 1983, 981–1000, 1033 ff.; for Welles' visit and his long conversation with Eden, see in particular National Archives (NA), UK: PREM 4/25/2, 13/3/1940. See also *F.J. Harbutt*, Churchill, Hopkins and the Other Americans: An Alternative Perspective on Anglo-American Relations, 1941–45, International History Review 8 (1986).

24 For this meeting, see *L.C. Gardner*, Spheres of Influence, 91 ff. (quote: 99); also *The Atlantic Charter*, ed. by D. Brinkley and D. R. Facey-Crowther, Basingstoke 1994; *T. A. Wilson*, The First Summit: Roosevelt and Churchill at Placentia Bay 1941, London 1969, rev. ed. 1991, 8 ff., esp. 166–67, 173 ff.; also *K. V. Morton*, Atlantic Meeting [. . .], London 1946; *R. Dallek*, Franklin D. Roosevelt and American Foreign Policy, 1932–1945, New York 1979, 281–86; *W. F. Kimball*, Anglo-American War Aims, 1941–43, The First Review: Eden's Mission to Washington, in: *The Rise and Fall of the Grand Alliance, 1941–45*, ed. by A. Lane and H. Temperley, Basingstoke 1995, esp. 1–3.

25 Point three of the Atlantic Charter's eight 'common principles' (and effectively four major goals) expressed 'the right of all peoples to choose the form of government under which they will live' and it endorsed the desire 'to see sovereign rights and self-government restored to those who have been forcibly denied them'. Point four

On 21 December 1941, two weeks after the Japanese attack on Pearl Harbor, Churchill met with Roosevelt for prolonged discussions in Washington. The Prime Minister had more or less invited himself and could not be dissuaded by news from Lord Halifax, the British ambassador, that the President had expressed reservations about the planned meeting. Roosevelt, however, succeeded in persuading Churchill not to rush to Washington immediately after Pearl Harbor[26]. With his first visit to Washington as Prime Minister Churchill was able to give significant impetus to the formation of the 'Grand Alliance' of the Second World War and he thereby laid the basis for victory over the axis powers and for the physical survival of Britain[27].

Churchill would never forget his highly successful attempts, from the middle of 1940 to the end of 1941 to bind the United States ever more closely to his country and obtain American support for the British war effort and, as he believed, for the preservation of both Britain and the Empire. After the Japanese attack on Pearl Harbor he exulted:

> *"I knew the United States was in the war, up to the neck and in to the death. So we had won after all! ... England would live; Britain would live; the Commonwealth of Nations and the Empire would live."*[28]

---

dealt with the right of all states to have equal access to trade and raw materials. Quoted in *Gardner*, ibid., 99. See *W. R. Louis*, American Anti-Colonialism and the Dissolution of the British Empire, in: *The 'Special Relationship': Anglo-American Relations since 1945*, ed. by W. R. Louis and Hedley Bull, Oxford 1986, 61–83; *W. R. Louis*, Imperialism at Bay. The United States and the Decolonization of the British Empire, 1941–45, New York 1978, esp. 3–26, 48–67, 121–33, 548–73; *Kimball* (note 22 – The Juggler), 127–52; *R. B. Woods*, A Changing of the Guard: Anglo-American Relations, 1941–46, Chapel Hill 1990, 43–62.

26 See NAUK: PREM 4/27/9, letter Halifax, Washington, to Churchill, 9/12/1941, reporting on his conversation with Roosevelt. Churchill's talks with FDR were continued in January by their respective military experts who began to discuss joint strategy; this laid the basis for the Anglo-American Combined Chiefs of Staff which was soon established. See *W. S. Churchill*, The Second World War, Vol. 3: The Grand Alliance, London 1950, 608–09; *J. Baylis*, Anglo-American Defence Relations, 1939–1980. The Special Relationship, New York 1981, 5–7.

27 See *K. Sainsbury*, Churchill and Roosevelt at War: the War they Fought and the Peace they hoped to Make, London 1994; *Allies at War: the Soviet, American, and British Experience, 1939–45*, ed. by D. Reynolds et al., New York 1994; *M.S. Stoler*, Allies and Adversaries: The Joint Chiefs of Staff, the Grand Alliance, and US Strategy in World War Two, Chapel Hill 2001.

28 Quoted in *Churchill*, Second World War, Vol.3, 539. For the controversy about whether or not Churchill passed on all information regarding the Japanese attack on Pearl Harbor, see *Kimball*, Wheel within a Wheel, 298; also *Kimball*, Forged in War, 120–23; see also the controversial and misleading book by *J. Rusbridger* and *E. Nave*, Betrayal at Pearl Harbor: How Churchill Lured Roosevelt into World War II, New York 1991.

After all, only a month prior to this event Roosevelt had doubted whether he would get a declaration of war from Congress if he asked for it; after Pearl Harbor this was no problem[29]. Although Churchill's emotional outburst ignored the very real threat after Pearl Harbor that the US might focus on the war in the Pacific rather than on the European theatre of war, it soon became clear that Churchill was essentially right. According to Lord Halifax the Roosevelt administration was "*terribly shaken*" by the Japanese assault, and

> "*fully realize that they have been caught napping. I think they realise too what it means.*"[30]

Close Anglo-American relations would indeed prove to be of monumental importance for the further course of the war and for the very survival of Britain and the western world as a whole. The Prime Minister believed that above all his personal diplomacy, including his correspondence with Roosevelt, frequently conducted by outflanking the Foreign Office, was the main reason for the success of his policies[31]. Subsequently, this experience of American support was complemented by many other examples of

---

29 NAUK: PREM 4/27/9, letter Halifax, Washington, to Churchill, 11/10/1941, reporting on his conversation with Roosevelt on 10/10/1941.

30 Ibid., letter Halifax, Washington, to Churchill, 9/12/1941, reporting on his conversation with Roosevelt. Yet, the Americans resented the fact that Churchill referred in a speech to America's entry into the war as something he had 'dreamt of, aimed at and worked for'. It sounded like a Macchiavellian Churchillian policy of getting 'American boys' into the war: 'first our arms, then our money, then our boys'. Ibid., letter Halifax to Eden, 24/2/1942. For the recent developments in the international history of World War II, see The Future of World War II Studies: A Roundtable, Diplomatic History 25/3 (2001), 347 ff.

31 Churchill fully believed Lord Halifax's assessment: 'I have no doubt that the more you can maintain your private communications with him and tell him everything that you can, the better. He continually gets … rather partial and limited impressions from many of their people, and if you could manage to give him from time to time the picture as you yourself see it, it would be an inevitable correction to other stuff that is pouring in on him all the time'. PRO: ibid., letter Halifax, Washington, to Churchill, 11/10/1941, reporting on his conversation with Roosevelt on 10/10/1941. See also *Churchill*, ibid., 207–09, 377, 380, 384, 608–09, 594–96. The closeness of the Churchill-Roosevelt relationship should however not be exaggerated. Despite relatively few initial differences of opinion about immediate war aims (long-term post-war aims were a very different matter) and the sharing of the atomic secret and both country's code-breaking capabilities, the image of intimacy and total candour in their relations with each other as portrayed by Churchill and many subsequent historians is misleading. For a good critical overview, see already *Sherwood*, Roosevelt and Hopkins; and *Kimball*, Wheel within a Wheel, 291 ff.; also *D. Reynolds*, Roosevelt, Churchill, and the Wartime Anglo-American Alliance, 1941–45: towards a New Synthesis, in: *Special Relationship*, ed. by Louis and Bull, 17–41; *P. Boyle*, The Special Relationship: An Alliance of Convenience?, Journal of American Studies

Washington's generosity which he secured through personal negotiations. For example, in June 1942 Churchill happened to be in Washington for his third visit as Prime Minister when news of the loss of the British garrison in Tobruk arrived; he was able to persuade Roosevelt to offer immediate replacements for the lost tanks and guns[32].

Moreover, during 1942–43 Churchill managed to obtain Roosevelt's support for the North African campaign and his Mediterranean strategy (invasions of Sicily and Italy) which necessitated a postponement of the opening of a Second Front in France, to alleviate the German pressure on the Soviet Union, until 1943 and ultimately even 1944. Although this was strongly opposed by the American Chiefs of Staff, including General Marshall, who favoured a rapid attack on Germany through France, the President gave in to Churchill's powers of persuasion[33].

During the entire war Churchill continued to seek to arrange meetings with Roosevelt on a regular basis and took care not to neglect his correspondence with the President. In total Churchill and Roosevelt met eleven times without Stalin, usually every four to five months[34]. Prior to all war summit conferences of the 'Big Three' at Tehran, Yalta and Potsdam, the Prime Minister suggested meetings with Roosevelt so as to identify a joint position vis-à-vis the USSR and increase the chance of impressing the British point of view on Roosevelt. He was however not always successful with this strategy[35].

Before the Tehran Conference in November/December 1943 Churchill was the only one of the 'Big Three' who had maintained personal, direct contact with the other two. He was extremely pleased with such an arrangement and cared little for the prospect of Roosevelt and Stalin assuming close bilateral relations. Churchill's strong anti-Communism made him deeply mistrust the Soviet dictator. For example, in 1942 and 1943 he frequently suspected Stalin of considering the conclusion of a separate peace treaty with Hitler[36]. To a considerable degree, however, Churchill was also distrustful

---

22 (1988), 457–65. For the sharing of intelligence information see, *B. Smith*, Sharing ULTRA in World War II, International Journal of Intelligence and Counterintelligence 2 (1988), 59–72.

32 See *Kimball*, Wheel within a Wheel, 298; *W. F. Kimball* (note 22 – The Juggler), 173–83.

33 For a succinct summary, see *F. Harbutt*, The Iron Curtain: Churchill, America, and the Origins of the Cold War, New York 1986, 47–48.

34 See *Kimball*, Wheels Within a Wheel, 296. Usually, however, Churchill had to travel to Washington or they met in third countries; FDR never came to London during the war.

35 See *Churchill & Roosevelt* (note 16), 4, 11–13, 15; *Barker*, Churchill and Eden, 130.

36 See for example NAUK: PREM 4/100/8, minute Eden to Churchill, P.M./43/198, 2/7/1943. See also *Weinberg*, World at Arms, 289, 462, 609–11; *I. Fleischhauer*,

of Roosevelt. For instance, he tremendously disliked the President's devious policy of ignoring Britain when attempting to obtain Stalin's co-operation for his internationalist design for the post-war world. Above all, Churchill suspected both Roosevelt and Stalin of intending to place Britain's far-flung colonial interests on the negotiating table and, in particular during the last two years of the war, ignore many of his country's strategic concerns and preferences for the shape of the post-war world[37].

Yet, despite all Anglo-American difficulties, the Prime Minister was always convinced that the 'fraternal association' with the United States, brought about by the common language and a similar democratic culture, was of a very different quality from the alliance with the Soviet Union[38]. Nonetheless, the latter was no less important to Britain's survival and the country's future role in the world. Thus, whilst Churchill attempted successfully to maintain close contacts with the American President, he was also greatly interested in pursuing a close relationship with Stalin. He sought to correspond with and meet the dictator as often as possible[39].

Stalin, however, remained deeply distrustful of Churchill and his long-standing anti-Communist reputation[40]. Illustrative of Stalin's great mistrust

Die Chance des Sonderfriedens: Deutsch-sowjetische Geheimgespräche 1941–1945, Berlin 1986. For Churchill's suspicion of Stalin, see concisely *Harbutt*, Iron Curtain, 23 ff., 67–68.

37 See *K. Sainsbury*, The Turning Point: the Cairo, Moscow and Teheran Conferences, London 1985, 9–10, 293 ff.; *G. L. Weinberg*, A World at Arms: A Global History of World War II, Cambridge 1994, 591 ff., 625 ff.

38 For the deeper roots and the gradual development of the so-called Anglo-American 'special relationship', see for example *H. C. Allen*, Great Britain and the United States. A History of Anglo-American Relations, 1783–1952, Watford, 1954; *D. C. Watt*, Succeeding John Bull: America in Britain's Place, 1900–1975 [...], Cambridge 1984; *The Special Relationship*, ed. by Louis and Bull; *Reynolds* (note 16); *An Ocean Apart: the Relationship between Britain and America in the 20th Century*, ed. by D. Dimbleby and D. Reynolds (London, 1988); *C. J. Bartlett*, 'The Special Relationship': A Political History of Anglo-American Relations since 1945, London 1992; *A. P. Dobson*, Anglo-American Relations in the Twentieth Century, London 1995.

39 See *Churchill & Roosevelt* (note 16), 9–10, 12–13. However, he met Stalin only twice without Roosevelt: in 1942 and 1944, both times in Moscow. Eden very much resented Churchill's tendency to monopolize contacts with both Roosevelt and Stalin. See *D. Dutton*, Anthony Eden: A Life and a Reputation, London 1997, 183. See also *S. M. Miner*, Between Churchill and Stalin: the Soviet Union, Great Britain and the Origins of the Grand Alliance, Chapel Hill 1988; also *R. Edmonds*, Churchill and Stalin, in: *Churchill*, ed. by Blake and Louis, 309–26. For revealing selections of documents, see *The Foreign Office and the Kremlin: British Documents on Anglo-Soviet Relations, 1941–45*, ed. by G. Ross, Cambridge 1984.

40 For some of the best overviews on Stalin's life and political views, see *E. Radzinsky*, Stalin, New York 1996, 495 ff.; *D. A Volkogonov*, Stalin: Triumph and Tragedy,

of Churchill are the words he apparently uttered to the visiting Yugoslav communist Djilas in 1944:

> *"Perhaps you think that just because we are the allies of the English we have forgotten who they are and who Churchill is ... Churchill is the kind of man who will pick your pocket of a kopeck if you don't watch him."*[41]

Since early 1942 Stalin had known of the secret Anglo-American research programme to develop an atomic bomb which he was not told about; together with the issue of the Second Front it decisively contributed to the dictator's continued great suspicion of the Anglo-Americans[42]. Moreover, the Soviets were aware of the fact that Churchill had never been able to overcome entirely his deeply ingrained and long-standing anti-Communism which went back as far as the year 1917.

Yet with respect to overcoming the Bolshevik rule in Russia, Churchill was not prepared to contemplate negotiations. Not unlike his contempt of Hitler and Nazism after 1938/39, Churchill regarded any attempt to arrive at a diplomatic solution with Lenin's butchers in Russia as beyond the pale. Before the very survival of Britain herself was threatened by Hitler's onslaught in 1940 he found it impossible to overcome his visceral anti-Communism.

It appears that it was only at some stage between 1933 and 1936 that Churchill became aware that the greatest danger to world peace was not coming from the Soviet Union but from Hitler's Germany. Thus, he gradually began to dampen his criticism of Stalin and to differentiate between Stalin as a person and international communism as such. Whilst he continued to regard the latter as an enemy which had to be fought, he expressed increasingly greater admiration for Stalin[43].

London 1991, 483–93; *I. Deutscher*, Stalin: A Political Biography, rev. ed., Harmondsworth 1990, 451 ff.

41 *M. Djilas*, Conversations with Stalin, London 1963, 61. See also Stalin's reference to Roosevelt and Churchill as 'comrades-in-arms' during the Tehran Conference. He had quickly added 'if it is possible for me to consider Mr Churchill my friend'. FRUS: The Conferences at Cairo and Tehran, 1943, 837: Memorandum Bohlen, Dec. 1943.

42 In August 1943 Churchill and Roosevelt agreed not to share the results of this research with anyone; Churchill explained that he regarded the atomic bomb as a weapon which could be used to contain Moscow after the end of the war. See *Kimball* (note 22 – The Juggler), 87; for an excellent account of the Soviet nuclear programe, see *D. Holloway*, Stalin and the Bomb: the Soviet Union and atomic energy, 1939–56, New Haven 1994.

43 See *Harbutt*, ibid., xii, 16–17, 31–32. For an overview of Churchill's statements on the Soviet Union between 1918 and 1940, see the collection of quotes and cartoons by *Mon Alié Staline de Winston Churchill*, ed. by Victor, Paris 1942; also *Sir Winston Churchill: A Self-Portrait [...]*, ed. by C. R. Coote, London 1954, 72–83.

In principle, before, during and also after the Second World War, Churchill remained loyal to his strong anti-Communist convictions; his deep suspicion of Lenin's successor in the Kremlin persisted. He would always be repelled by the general egalitarian, anti-monarchical, anti-democratic and atheistic nature of Communism. Churchill was however fully aware that it had been Stalin who had offered London and Paris a tripartite alliance in May 1939 which against Churchill's strong advice had not been accepted by Prime Minister Chamberlain and Foreign Secretary Halifax. Churchill regarded the Soviet proposal as quite 'a fair offer' and, in his memoirs, he referred to Stalin's subsequent treaty with Hitler as a 'realistic' decision[44].

Yet, despite a certain degree of understanding for Stalin's action from a real-political point of view, Churchill was never able to overcome his resentment of the fact that the USSR had concluded a neutrality pact with Nazi Germany in 1939. Only Hitler's invasion of the Soviet Union had brought Stalin onto the side of the western democracies. Churchill was firmly convinced that if Britain

> "*had been invaded and destroyed in July 1940 or August 1940 ... they* [the Soviets] *would have remained entirely indifferent*"[45].

Churchill's attitude towards co-operation with the Soviet Union is well expressed in his famous statement:

> "*I have only one purpose, the destruction of Hitler ... If Hitler invaded Hell I would make at least a favourable reference to the Devil in the House of Commons.*"[46]

Thus, whilst he remained ideologically deeply hostile to Communism, during the Second World War he was very much alive to the fact that Britain needed Stalin to defeat Hitler. Naturally, despite Churchill's changing rhetoric, Stalin and his lieutenants were fully aware of Churchill's feelings about their regime.

---

44 Quoted in *Edmonds*, Churchill and Stalin, 312; see also *W. S. Churchill*, The Second World War, Vol. 1: The Gathering Storm, rev. ed., London 1949, 289–308; *D. C. Watt*, How War Came, London 1990, 370–72; *C. Thorne*, The Approach of War, 1938–39, London 1967, 140–47; *G. Niedhart*, Großbritannien und die Sowjetunion, 1934–39 [...], Munich 1972, 390–425.

45 Churchill to Cripps, the British Ambassador in Moscow, 28/10/1941. Quoted in *W. S. Churchill*, The Second World War, Vol. 4: The Hinge of Fate, London 1951, 443.

46 *Churchill*, Second World War, Vol. 3, 331. For Churchill's similar explanation for his change in view given to Soviet ambassador Ivan Maisky in 1938, see *Edmonds*, Churchill and Stalin, 311.

As early as July 1940, long before the German invasion of the Soviet Union, Churchill highlighted the common threat that the possible domination of Nazi Germany would pose to the European continent. He sent a personal letter to Stalin and suggested Anglo-Soviet consultations. The dictator did not respond favourably[47]. Based on Britain's ability to break the Enigma code of the encrypted secret German radio messages, in April 1941 Churchill warned Stalin about the imminent German invasion of the Soviet Union. However, Stalin refused to believe that something like Operation BARBAROSSA was possible. He entirely misjudged Hitler and ignored Churchill's and British ambassador Stafford Cripps's warnings as well as the numerous other indications from his intelligence agents of an impending German invasion.[48] Stalin even assumed that these were British ploys to provoke Moscow into fighting Germany. Soviet mistrust was particularly aroused by the flight of Rudolf Hess to Scotland in May 1941, which was regarded in Moscow as the prelude to British-German peace talks. This episode encouraged Stalin to discard all British information about Hitler's imminent invasion of the Soviet Union[49].

Based on his Marxist interpretation of the direction of the immanent forces of history, Stalin continued to believe that the capitalists and western

---

47 See *Harbutt*, Iron Curtain, 33; also for a good concise overview of Anglo-Soviet relations during the war, *V. Rothwell*, Britain and the Cold War, 1941–47, London 1982, 74 ff., 151 ff.; and the good detailed account by *M. Kitchen*, British Policy towards the Soviet Union during the Second World War, Basingstoke 1986; see also *G. Warner*, From ally to enemy: Britain's relations with the Soviet Union, 1941–48, in: *Diplomacy and World Power*, ed. by Dockrill and McKercher, 221–43.

48 See above all *G. Gorodetsky*, Grand Delusion: Stalin and the German Invasion of Russia, New Haven 1999, Ch.8; see also the excellent review by R. Overy in the TLS, 11 August 2000. See also *Gardner*, Spheres of Influence, 86–87; *Stafford*, Secret Service, 221–23; *O. Ordietsky*, Churchill's Warning to Stalin: a reappraisal, Historical Journal 29/4 (1986), 979–90; also *Molotov Remembers: Inside Kremlin Politics. Conversations with Felix Chuev*, hg. v. A. Resis, Chicago 1993, 28. There has still not been any evidence found for the old theory that Hitler's invasion was a preventive war as Stalin was mobilizing for an immediate attack on Germany. See *Gorodetsky*, ibid., and *Präventivkrieg? Der deutsche Angriff auf die Sowjetunion*, ed. by B. Pietrow-Ennker, Frankfurt/M. 2000.

49 See *Gorodetsky*, ibid., Ch.12, esp. 267–74; *I. Kershaw*, Hitler, 1936–1945: Nemesis, London 2000, 369 ff.; *P. Padfield*, Hess: Flight for the Führer, new ed., London 2001; also *R. F. Schmidt*, Rudolf Hess, 'Botengang eines Toren?' Der Flug nach Grossbritannien vom 10. Mai 1941, Düsseldorf, 1997; and for the confusion in London caused by Hess's flight, *R. F. Schmidt*, Der Hess-Flug und das Kabinett Churchill. Hitlers Stellvertreter im Kalkül der britischen Kriegsdiplomatie Mai-Juni 1941, Vierteljahrshefte für Zeitgeschichte 42/1 (1994), 1–38. See also *Martin*, Friedensinitiativen und Machtpolitik im Zweiten Weltkrieg, 1939–1942, Düsseldorf 1974, 425–47; *U. Schlie*, Kein Friede mit Deutschland: Die geheimen Gespräche im Zweiten Weltkrieg, 1939–1941, Munich 1994, 290–324; also *Costello*, Ten Days to Destiny.

imperialists would fight it out among themselves and that, contrary to all the available evidence, Hitler intended to defeat Britain first before turning towards the East. Stalin was convinced that Germany wanted to avoid a war on two fronts and therefore assumed that Hitler would negotiate or at least send an ultimatum before embarking on a war against the Soviet Union[50].

However, on 22 June 1941 Hitler invaded the Soviet Union and Churchill publicly welcomed the Soviet Union as a new ally and promised immediate British help to the Russian nation[51]. Although British-Soviet suspicions remained high, on July 8 and 10 two more messages from Churchill were delivered to Stalin. They resulted in Anglo-Soviet negotiations and the eventual signing of an Anglo-Soviet Declaration in Moscow and the mutual promise not to enter into separate peace agreements with Hitler[52]. Soon, however, Churchill's refusal to give a commitment to open a Second Front in France in the near future to relieve the pressure of the invading German forces on the Red Army aroused Stalin's anger and increased his great suspicion of British motives[53]. The Prime Minister believed that

> "*there was no point in opening another front or shedding blood just to please the Russians*".

After all, Stalin himself had only entered the war "*to protect their own interests*"[54]. Churchill was therefore not moved by Stalin's anger; he realized that the dictator was attempting to imbue him with a sense of guilt by repeatedly emphasizing the heavy toll of casualties the Soviet army (and indeed the

---

50 Ibid.

51 On the day after the German invasion, Churchill broadcast: 'No one has ever been a more consistent opponent to Communism than I have for the last twenty-five years. I will unsay no word that I have spoken about it. But all this fades away before the spectacle which is now unfolding. We shall give whatever help we can to Russia and the Russian people.... The Russian danger is therefore our danger, and the danger of the United States, just as the cause of any Russian fighting for his hearth and home is the cause of free men and free peoples in every quarter of the globe ...'. Quoted in *Edmonds*, Churchill and Stalin, 313.

52 For the negotiations, see the documents in *War and Diplomacy: The Making of the Grand Alliance. Documents from Stalin's Archives*, ed. by O. A. Rzheshevsky, Amsterdam 1996, Part I, 1–62; for a summary on the history of British-Soviet relations, see also *The Foreign Office and the Kremlin*, ed. by Ross, 3–9, 10 ff.

53 See for example *J. Haslam*, Soviet War Aims, in: *The Grand Alliance* (note 24), 36–37; also *J. Erickson*, Stalin, Soviet Strategy and the Grand Alliance, in ibid., 136–141; *R. Edmonds*, The Big Three: Churchill, Roosevelt and Stalin in Peace and War, London 1991, 282 ff.

54 NAUK: PREM 3/366/8, W.P.(43)430, 5/10/1943: Record of a meeting between Churchill and Dominion Representatives in Washington held at the British Embassy, Washington, on 10th September 1943.

civilian population) was suffering. The Prime Minister informed the British ambassador in Moscow that the Soviets appeared to assume

> *"that they were conferring a great favour on us by fighting in their own country for their own lives. The more they fought the heavier our debt became."*

Churchill concluded that this "*was not a balanced view*"[55].

In July and September 1941 Stalin asked for military aid from Britain. In view of the German armies' push towards Moscow, the Russian dictator hinted in his message to Churchill of 8 November at his readiness to participate in an agreement on war objectives and the organisation of the post-war world[56]. Despite the Prime Minister's resentment at the tone of Stalin's letter, in December he eventually sent Foreign Secretary Anthony Eden to the Soviet capital. Not unlike Churchill, Eden also believed strongly in his own ability to overcome difficult situations by means of personal diplomacy[57]. However, due to disagreements regarding the future frontiers of the post-war world and Soviet insistence on the recognition of its 1941 borders as negotiated with Hitler, the envisaged Anglo-Soviet Treaty of Alliance with a duration of 20 years could not be sealed[58]. The treaty was only signed with Soviet Foreign Minister Molotov in London in late May 1942[59].

---

55 Quoted in *Harbut*, Iron Curtain, 46. Churchill's letter was dated 28/10/1941.

56 *Correspondence between the Chairman of the Council of Ministers of the USSR and the Presidents of the USA and the Prime Ministers of Great Britain during the Great Patriotic War of 1941–45*, ed. by Ministry of Foreign Affairs of the USSR, Moscow 1957, 33–34 (Doc. No.20), and for Churchill's reply, 34–35 (Doc. No.21).

57 He told Polish Prime Minister in exile, General Sikorski, that his government 'had such evidence of the deep suspicion which had implanted itself in the Soviet Government's mind that there was not alternative but for him to go'. Quoted in *Rothwell*, Britain and the Cold War, 85.

58 PRO: PREM 4/30/8, Record of Interview Between the Foreign Secretary and M. Stalin, December 16, 1941. See also the following note.

59 Eden had also attempted to obtain Molotov's agreement to his idea that both London and Moscow would not conclude any bilateral treaties with the Eastern European countries as this would only lead to a competition for allies. On Eden's talks in Moscow and Molotov's negotiations London, see the Soviet documents in *War and Diplomacy*, ed. by Rzheshevsky, Parts I, II and IV; *A. Eden (Lord Avon)*, The Reckoning, London 1965, 275 ff., esp. 289–303, 327–30; also *Dutton*, Anthony Eden, 185–96; *V. Mastny*, Russia's Road to the Cold War: Diplomacy, Warfare and the Politics of Communism, 1941–45, New York 1979, 41–45; also *R. Overy*, Russia's War, London 1998. It was Eden and not so much Churchill who argued particularly strongly in favour of aid to the Soviet Union. See *S. Lawlor*, Britain and the Russian Entry into the War, in: *Diplomacy and Intelligence during the Second World War*, ed. by R. Langhorne, Cambridge 1985, 168–83. For the thinking of the FO, see also *G. Ross*, Foreign Office Attitudes to the Soviet Union, 1941–45, Journal of Contemporary History 16 (1981), 522–24.

Together with Roosevelt's confidant Averell Harriman, the Prime Minister himself went to Moscow in August 1942 to inform Stalin in person that the envisaged but constantly postponed Second Front in northern France could still not be opened by the end of the year[60]. Churchill and Stalin met four times and in spite of one very difficult and vociferous session with the dictator, he felt 'encouraged' by his visit. Stalin had after all informed him that "*we like a downright enemy better than a pretending friend*"[61]. The Prime Minister told Roosevelt later that the atmosphere in the Kremlin was generally very "*easy and friendly*" and that he had been able to build up a "*personal relationship*" with Stalin. He also reported to the President that

> "*the disappointing news I brought could not have been imparted except by me personally without leading to really serious drifting apart*"[62].

Churchill was not prepared to leave discussions with Stalin or Roosevelt in the hands of the diplomats of the British Foreign Office. Although his officials were mostly quite competent, in Churchill's view they lacked imagination and flair. Even Foreign Secretary Eden would only be sent on reconnaissance missions in the event that the Prime Minister was unable to go himself or if it was inappropriate for the head of government to negotiate in person; thus, not Churchill but Eden attended the foreign ministers' conference in Moscow in October 1943[63]. The drawback of this kind of diplomacy lay in the fact that Churchill's sentimental and romantic personality and his very subjective and often idiosyncratic preferences and dislikes gained an importance they would not have achieved if the Prime Minister had relied

---

60 For Churchill's meetings with Stalin and the reports and telegrams he sent to Roosevelt and British Deputy Prime Minister Clement Attlee, see *Roosevelt & Churchill* (note 16), 560–72; also *Kitchen*, British Policy, 124 ff., esp. 132 ff. Operation Overlord, the Normandy invasion, only took place in February 1944. In the meantime Churchill endeavoured to convince Washington that it was necessary to invade the Balkans and strengthen the Mediterranean militarily, a policy which above all served to strengthen Britain's great power role in this region and the Near East. See *W. Loth*, Die Teilung der Welt, 1941–1955, Munich 1980, 51 f., 71–75, esp. 72. See also *T. Ben-Moshe*, Winston Churchill and the "Second Front": A Reappraisal, Journal of Modern History 62 (1990), 503–37; also *M. A Stoler*, The Politics of the Second Front: American military planning and diplomacy in coalition warfare, 1941–43, Westport, Con. 1977; Kimball (ed.), ibid., 6, 13; *R. Douglas*, From War to Cold War, New York 1981, 9–11.

61 Quoted in *Churchill*, Second World War, Vol. 4, 443.

62 Quoted in *Churchill & Roosevelt* (note 16), 571–72: Churchill to Roosevelt, 18.8.1942. See also *Churchill*, ibid., 428–50; *Harbutt*, Iron Curtain, 47; *Barker*, Churchill and Eden, 224.

63 See *Barker*, ibid., esp. 15–28, 125–147; *Churchill & Roosevelt: The Complete Correspondence, Vol. 3: Alliance Declining*, ed. by W. F. Kimball, London 1984, 180: Churchill to Roosevelt, 11.6.1944.

to a greater degree on the more restrained advice and sober judgement of his official advisers. Therefore, it may well be correct to assume that, for example, in his 1942 meetings with Stalin in Moscow, as with Roosevelt at Placentia Bay the year before, "*Churchill probably persuaded himself more effectively than he did his interlocutor*" of the correctness and persuasiveness of his opinions[64]. Moreover, in the course of his meetings and his correspondence with Stalin, Churchill found it increasingly difficult to escape the dictator's charisma[65].

Churchill (as well as Roosevelt) was therefore not immune to the myth which developed during the war that Stalin himself was quite a benign person with a common sense approach but that he was directed by evil men in the background who, perhaps, belonged to the military or the security services which were overseen by the evil Beria. Thus, it was concluded in many quarters in both Britain and the United States that Stalin could not act as he would have liked[66]. This was of course mere wishful thinking which, however, lingered on into the post-war period[67]. During the war conferences of the 'Big Three' from Tehran to Potsdam, as well as during Churchill's bilateral talks with Stalin in 1942 and in October 1944 (and in the course of casual conversations during the war conferences), the Prime Minister was convinced that only he personally could obtain concessions from Stalin and move the dictator towards a more compromising mode of co-operation with the Anglo-Saxon powers.

Churchill's personal diplomacy met with fierce competition from Roosevelt, the American President who, in 1933, had first entered into diplomatic relations with the USSR[68]. As early as July 1941, even before America's en-

64 *Edmonds*, Churchill and Stalin, 316. On Churchill's romantic vision in foreign affairs, see also *I. Berlin*, Churchill in 1940, 18.

65 Not without reason did Stalin receive the pet name 'Uncle Joe' in Britain and America. Hardly anyone would have dreamed of inventing a similar nickname for Soviet Foreign Minister Molotov. Allegedly Stalin was not too pleased about this nickname. See *Churchill & Roosevelt* (note 63), 573; *Churchill*, The Second World War, Vol. 6: Triumph and Tragedy, London 1954, 345.

66 See for example *Churchill*, Second World War, Vol. 4, 441; NAUK: PREM 4/100/7, PM's Personal Minute M.474/2 to Eden, 21/10/1942; *Snow*, Variety of Men, 213–14; *Harbutt*, Iron Curtain, 47. During the war people like for example Ernest Bevin and George Kennan shared this view at times. See *Barker*, Churchill and Eden, 221–22, 224.

67 Western illusions and Stalin's charisma were so powerful that even after the end of the Second World War American presidents Truman and Eisenhower were on occasion and despite clear evidence to the contrary influenced by this myth. See for example the memorandum of the 136th NSC meeting, in *FRUS* 1952–54, Vol.8, 1118, 11/3/1953.

68 For Roosevelt's policy towards Stalin, see for example *Harbutt*, Iron Curtain, 38–45; *Sainsbury*, Turning Point, 7–10; *J. L. Gaddis*, The United States and the Origins of

try into the war and five months before Eden's first wartime trip to Moscow, Roosevelt had already sent his confidant Harry Hopkins to the Soviet capital with the task of beginning consultations over extending the Lend-Lease programme to the Soviet Union[69]. Roosevelt also strongly believed, as he informed Churchill in "*brutally frank*" terms in February 1942, that he himself was able to

> "*handle Stalin better than either your Foreign Office or my State Department. Stalin hates the guts of all your top people. He thinks he likes me better, and I hope he will continue to do so.*"[70]

In May 1943 the President's special ambassador, the pro-Soviet Joseph Davies, travelled to Russia to convey proposals for a secret round of talks between Stalin and Roosevelt. Above all, the President planned to use such a meeting to discuss the question of a Second Front whose constant postponement aroused greater mistrust in Stalin regarding the intentions of the Anglo-Saxon powers than any other problem. It appeared to the President that only a personal clarification of the situation could bring about long-term trustful co-operation with the USSR[71].

Roosevelt mentioned nothing of his plans to Churchill, who still regarded the idea of a Second Front in France at this time as premature. The Prime Minister preferred to focus on the Mediterranean theatre of war to secure the lines of communication with Britain's colonial possessions and protect the British Empire. In his correspondence with Churchill Roosevelt even denied his intention to meet Stalin alone and thereby deliberately deceived

---

the Cold War, 1941–1947, New York 1972, 4 ff., 32 ff.; also *Loth*, Teilung der Welt, 27 ff.; *J. L. Gaddis*, Russia, the Soviet Union and the United States. An Interpretive History, New York 1978, 147 ff.; *R. Edmonds*, The Big Three: Churchill, Roosevelt and Stalin in Peace and War, London 1992, 363 ff.; *Kimball* (note 22 – The Juggler), 159 ff.; also *A. Perlmutter*, FDR & Stalin: a not so grand alliance, 1943–45, Columbia, Mo. 1993.

69 After the German invasion of the Soviet Union American public opinion became noticeably more sympathetic to Russia. See *R. B. Levering*, American Opinion and the Russian Alliance, 1939–1945, Chapel Hill, N.C. 1976, 43–62. Regarding 'lend-lease' aid for Moscow see *Sherwood*, Roosevelt and Hopkins, 318–22, 323 ff.; *G. C. Herring*, Aid to Russia, 1941–46: Strategy, Diplomacy, and the Origins of the Cold War, New York 1973, 8 ff.; *L. Martel*, Lend-lease: Loans and the Coming of the Cold War. A Study of the Implementation of Foreign Policy, Boulder, Co. 1979, 25–56. For FDR's 'grand design' and long-term collaboration with Moscow, see *D. Yergin*, Shattered Peace: the Origins of the Cold War and the National Security State, Boston 1977, 55–58; *Loth*, Teilung der Welt, 40–41.

70 Quoted in *Churchill & Roosevelt* (note 16), 421: Roosevelt to Churchill, 18/2/1942.

71 For Davies' convictions, see *E. K. MacLean*, Joseph E. Davies: envoy to the Soviets, Westport, Conn. 1992, and *D. J. Dunn*, Caught between Roosevelt and Stalin: America's Ambassadors to Moscow, Lexington 1998, 61–94.

the Prime Minister. The American President did not believe that he needed Churchill to mediate between Washington and Moscow[72]. Much to the annoyance of the British Foreign Office, Roosevelt had already begun corresponding with Stalin unilaterally in February 1942. The President believed this would help him to obtain a better deal with Stalin regarding the future of the Baltic States. But to London it looked as if Washington did not seem to treat Britain as an equal partner. After all, contrary to the British practice, Washington hardly ever informed London about the bilateral exchanges of view with the Soviets although these conversations dealt with the Second Front and other questions of vital interest to Britain[73].

The President also avoided meeting the Prime Minister before the Tehran conference in late November 1943 when Roosevelt was to see Stalin for the first time. He did not wish to create the impression that the Anglo-Saxon powers were 'ganging up' on Stalin. During the Tehran meeting Roosevelt repeatedly turned down invitations from Churchill to have a personal chat. While thus refusing to talk in a more intimate way to his British friend, the President, however, did not hesitate to have three confidential talks with Stalin in the course of the conference[74].

It is clear that Churchill and Roosevelt were highly competitive, and even jealous of each other, in their mutual attempts to win the favour of Stalin. While Roosevelt did not entirely trust the British and their war aims and also felt that the predominant western power should not have to wait for permission from its much smaller ally before entering into contact with Stalin, Churchill realized that he was battling for Britain's survival as a great power. Thus, under all circumstances the Prime Minister intended to avoid being excluded from negotiations between Washington and Moscow. Just as was the case in his later strategy towards Presidents Truman and Eisenhower, the British Prime Minister constantly sought to check even the remotest tendency in Washington to call a bilateral American-Soviet summit conference

---

72 See *Churchill & Roosevelt* (note 16), 12–14; and the correspondence between the two leaders in ibid., Vol. 2, 233–34, 245–46, 259–61, 278–79, 283–84. See also *Sherwood*, Roosevelt and Hopkins, 733–34; *Loth*, Teilung der Welt, 71–74; *Sainsbury*, Turning Point, 8–11. Apart from the decreasing importance which Britain was to have in the post-war world which could already be recognised by 1943, Washington was also concerned about Churchill's old-fashioned Victorian views which seemed to pose a problem for the progressive restructuring of the post-war world. See *Kimball*, Wheel within a Wheel, 299, also 304–05.

73 See *Dutton*, Anthony Eden, 154.

74 See *FRUS*, Conferences at Cairo and Tehran, 1943, 482–86, 529–33, 594–96; also for example *Sainsbury*, The Turning Point, 218 ff.; *McNeill*, America, Britain and Russia, 355–57, 364–65, but see 368 ff. for their greater understanding during the second Cairo conference in December 1943.

which excluded Britain and might result in important decisions without the British government being able to influence them. Churchill was adamant in his determination to remain an equal partner of the 'Big Three'. He wished to be regarded as such not only by the other two powers but also in the general perception of the public[75]. It was only in the summer of 1943, when he was at his wit's end and when Britain's relations with Stalin had reached rock bottom, that Churchill spoke in support of Roosevelt arranging to meet 'Uncle Joe' alone[76]. Yet, such a meeting was never to take place.

During 1943 the Prime Minister became very much aware of the fact that his country had increasingly less influence on the planning and execution of allied war strategy. For example, in the spring of 1943 British-Soviet relations over yet another postponement of the Second Front had become very tense. After Stalingrad, when in one of the most ruthless and decisive battles of the war the Red Army managed to repulse the German onslaught on the USSR in late 1942 and early 1943 and thus effectively demonstrated that Hitler was no longer able to win World War II, Soviet military strength and indeed confidence was in the ascendant. It became clear that Stalin needed the help of Britain much less than hitherto. Even the Second Front and thus Moscow's desired assault on Germany through France was not regarded as quite as urgent as it had been previously. Furthermore, British support for ensuring the continued flow of provisions to the Soviet Union was no longer quite as vital as American goods were now able to reach Russia through Persia[77].

Churchill hoped that with a joint Anglo-American policy towards Stalin he would be able to balance his country's precarious position. He believed

> "*that we should do much better with the Russians if we first got on to intimate terms with the United States*".

It certainly

> "*was important not* [to] *allow the Russians to try to play the United States and the United Kingdom off against each other*"[78].

---

75 See *Harbutt*, Iron Curtain, 45, 49; *Dallek*, Franklin D. Roosevelt, 403–05.

76 Ibid.

77 See *Harbutt*, Iron Curtain, 49. For a good analysis of why the allies were eventually able to defeat Germany, see *R. Overy*, Why the Allies Won, pb. ed., London 1996, esp. 63 ff.; and for Russia's role, see in particular the same author's book Russia's War, Harmondsworth 1997, esp. 154 ff. A very popular recent account of the battle of Stalingrad is *A. Beevor*, Stalingrad, London 1998.

78 NAUK: PREM 3/366/8, W.P.(43)430, 5/10/1943: Record of a meeting between Churchill and Dominion Representatives in Washington held at the British Embassy, Washington, on 10th September 1943.

Yet Roosevelt, antagonized by Churchill's alleged imperialist ambitions, was fully aware of Britain's weakness and the Soviet Union's increasing strength. The President had little desire to form a united front with the British. Instead, in late 1943 Roosevelt posted the influential ambassador Harriman from London to Moscow, a move which is often interpreted as a rearrangement of priorities by the American government[79].

Britain's waning influence on war strategy became particularly obvious to Churchill during the Tehran Conference in November/December 1943, arguably one of the most important meetings of the war[80]. After all, during this somewhat chaotic and ill-prepared first 'Big Three' conference important deliberations took place regarding the timing of Operation OVERLORD (the new name for the Second Front strategy) during which the British preferences were largely ignored. Churchill also had to give in with regard to his post-war plans; instead, the nature and frontiers of the post-war world were almost entirely decided by Roosevelt and Stalin.

At Tehran Roosevelt and Stalin effectively entered into an implicit almost Faustian pact largely at the expense of their British ally. Stalin agreed to commit the USSR to participation in the American President's envisaged United Nations organization, which was to be organized on a global rather than a regional basis as Churchill desired. Roosevelt did not wish to antagonize American public opinion by being seen to accept any long-lasting entangling commitments in Europe. The dictator also promised to join the war in the Pacific against Japan once Hitler had been defeated. Both objectives were very dear to Roosevelt's heart. In return, the President appeared to offer Stalin the promise of American support for Moscow's envisaged territorial policy in Eastern Europe, in particular with regard to the Baltic countries and Poland. Above all, much to Churchill's displeasure, Roosevelt confirmed the American intention to commence operation OVERLORD as quickly as possible in 1944. He thus effectively agreed to reject Churchill's preferred strategy of assault on Germany from the Mediterranean, encouragement of Turkish entry into the fight against Hitler and increased resources for the war in Italy. Moreover, the President even wanted to begin with supporting operations for OVERLORD in the south of France

79 See for example *Harbutt*, ibid., 50; *Dallek*, Roosevelt, 410–11.

80 For a good and succinct overview, see ibid., 53 ff. See also the somewhat simplified account by *K. Eubank*, The Summit Conferences, 1919–1960, Norman, Okla. 1966, esp. 55–57, 58 ff. For excellent overviews of the conference and the developments leading up to it, see *Mastny*, Russia's Road, 111–44; *Sainsbury*, Turning Point, 12 ff.; *Weinberg*, World at Arms, 619–31.

(ANVIL) which would have led to the almost immediate termination of the Italian campaign[81].

Churchill, however, as has been argued, unwisely continued to focus on military rather than strategic issues in the course of the Tehran conference. It therefore became increasingly clear that in view of Roosevelt's and Stalin collusion, the Prime Minister occupied an ever more isolated position at Tehran[82]. To add insult to injury, Britain's role as a full member of the 'Big Three' club was also increasingly questioned by Stalin's and Roosevelt's dismissive attitude and behaviour towards Churchill. Not surprisingly, the Prime Minister found it difficult to overlook the condescension with which he personally was treated by both leaders. Although a few months before the conference he had uttered the view that he "*was always prepared to be snubbed for his country*"[83], making such an experience at the Tehran conference was a very different matter. Churchill was clearly most disconcerted by the unexpected and hurtful developments in the course of the conference[84].

On the whole, the Tehran meeting proved to be a disaster for Churchill and his country. The Prime Minister realized that after Tehran the major decisions concerning the remainder of the war and the post-war world would effectively be taken by the much more powerful Americans and Russians[85]. He later remarked in despair that he noticed "*at Tehran for the first time what a small nation we are*"[86]. Yet, in characteristic fashion Churchill re-

---

81 See *FRUS*, Conferences at Cairo and Tehran, 1943, esp. 487 ff., 529–552, 565 ff., 576 ff.; see also the literature in note 101 above, and in particular for a convincing interpretation *Harbut*, Iron Curtain, 53–60.

82 This is for example Fraser Harbutt's position. He also argues convincingly (ibid., p. 59) that while Roosevelt informed Stalin that he intended to demolish the British Empire after the war, Churchill attempted to ingratiate himself with Stalin and 'scramble back to the summit' at the expense of Poland and Turkey for which Britain had assumed treaty responsibilities.

83 PRO: PREM 3/366/8, W.P.(43)430, 5/10/1943: Record of a meeting between Churchill and Dominion Representatives in Washington held at the British Embassy, Washington, on 10th September 1943.

84 See for example Stalin's extraordinary political and personal attacks on Churchill during a tripartite dinner meeting on 29/11/1943: *FRUS*, Conferences at Cairo and Tehran, 1943, 553–55.

85 See *A. Hillgruber*, Der Zweite Weltkrieg, 124–28, 157; also *Sainsbury*, Turning Point, 293 ff.; *Barker*, Churchill and Eden, 28. Warren Kimball convincingly differentiates among four stages in the Churchill-Roosevelt relationship: 1. up to June 1940: the 'get acquainted stage'; 2. up to Pearl Harbor: Churchill was wooing Roosevelt to become an ally; 3. from 1942 to mid/late 1943: equal partners' stage; 4. after Tehran at the latest: junior-senior partner relationship existed. See *Kimball*, Wheel within a Wheel, 305.

86 Churchill said this to Violet Bonham-Carter and made a similar statement to the Czechoslovakian President Benes in February 1945. Quoted in Colville's essay in

mained deeply convinced of the superiority of his personal war and post-war strategy for both Germany's ultimate defeat and Britain's long-term survival as a great power. The Prime Minister succinctly summarised the situation at Tehran:

> *"There I sat with the great Russian bear on one side of me with paws outstretched, and, on the other side, the great American buffalo. Between the two sat the poor little English donkey, who was the only one who knew the right way home."*[87]

Churchill was deeply depressed by this state of affairs and attempted everything in his power to rectify the situation in the following months by intensifying his personal relations with Stalin and, above all, with Roosevelt. However, after Tehran the American President as well as his close associates became more aloof in their dealings with Churchill while also making increasingly more demanding requests on Britain's financial and economic resources; they also were less restrained in attempting to restrict the country's political and military independence of action. Indeed, as Fraser Harbutt has observed,

> *"Britain was being nudged, gently but firmly, toward second-rank status".*[88]

Yet, Churchill was not prepared to accept this. He hoped that above all with the help of his personal charisma he would eventually be able to move Roosevelt and even Stalin towards a war strategy and a post-war world more conducive to the interests and needs of Great Britain and its Empire. Churchill certainly had no intention of abandoning Britain's claim to great power status after Tehran. After all, fundamentally he was and remained one of Britain's most imperialistically orientated defenders of 'power politics'; he deeply believed in the continued importance of Britain and its Empire for the creation of a peaceful world. For example, he could never really accept the strong stirrings for independence in India and Egypt which became obvious during the war. His fierce opposition to offering Dominion status to

---

Action this Day: Working with Churchill, ed. by J. Wheeler-Bennett, London 1984, 96, n.1; also *Edmonds*, The Big Three, 318. During the first Quebec conference (14–28.8.1943) the United States had first insisted on their view in an important military-strategic decision against British doubts. Since their victory at Stalingrad in Feb. 1943 the Soviet Union's influence made itself also increasingly felt. See for example *A. Hillgruber*, Der Zweite Weltkrieg: Kriegsziele und Strategie der großen Mächte, 2nd ed., Stuttgart 1983, 125–26.

87 Quoted in *A. J. P. Taylor*, The War Lords, Harmondsworth 1978, 93–94.

88 *Harbutt*, Iron Curtain, 62.

India had made him resign from the Conservative shadow cabinet in January 1931, a decision which had cost him dearly as it had left him in the political 'wilderness' until 1939.

> "*The loss of India,*" he believed not altogether wrongly, "*would mark and consummate the downfall of the British Empire*"[89].

On the whole, during the Second World War he always clung to his stubborn belief that

> "*I have not become the King's First Minister in order to preside over the liquidation of the British Empire*"[90].
>
> Without doubt, Churchill possessed
>
> "*a romantic devotion to British greatness*" and "*a pugnacious determination to maintain the greatness of the British Empire*"[91].

It was his desire to maintain Britain's position as a global power as well as his wish to play a significant personal role on the global stage which inspired Churchill to insist on Britain's inclusion as one of the 'Big Three' and oppose vehemently Roosevelt's and Stalin's tendency to dismiss the Empire as a useless relic of the past, downgrade Britain's international importance and move towards a 'Big Two' alliance. Churchill was not prepared to accept the diminishing significance of his country in global politics as a permanent feature.

Once his preferred strategies for the continuation of the war and his vision for the post-war world had been rejected by Roosevelt and Stalin at Tehran, Churchill attempted to fall back on his summit diplomacy and use this method to prevent the further marginalization of Britain by the overwhelming power of the United States and the Soviet Union. The Prime Minister had realised that Britain, as by far the weakest member of the 'grand alliance', needed to reassert its continued membership of the 'Big Three' by

---

89 Quoted in *Rose*, Churchill, 189, see also 216 ff.

90 Churchill made this remark on 10 Nov. 1942 in the course of a speech during the Lord Mayor's Luncheon in London. Quoted in *Louis*, Imperialism at Bay, 200.

91 Quoted in *Taylor*, The Statesman, 22, 18. See also *Hillgruber*, Der Zweite Weltkrieg, 156–57; and Churchill's remarks in the chapter 'On Britain and the Empire,' in *Sir Winston Churchill. A Self-Portrait*, ed. by C. R. Coote, London 1954, 212–35, on India, ibid., 240–46. For Churchill's views on the British empire, see also *R. A. Callahan*, Churchill: Retreat from Empire, Wilmington, Del. 1984, 249; and above all *K. Emmert*, Winston S. Churchill on Empire, Durham, N.C. 1989; *R. Hyam*, Churchill and the British Empire, in: *Churchill*, ed. by Blake and Louis, 167–85; and *S. Gopal*, 'Churchill and India,' in ibid., 457–71.

all means possible. For this reason Churchill employed his formidable reputation, his long political experience and his great powers of persuasion to the utmost of his ability at the war conferences and during bilateral summit meetings to emphasise his country's view. He attempted to employ the whole force of his personality to Britain's advantage. Similar to his policy before the outbreak of the First World War when Churchill had sought to use summit diplomacy, "*to mitigate asperity between the German and British Empires*"[92], in particular during 1940–42 and then again in 1944–45 he attempted to use this strategy in his discussions with Stalin and Roosevelt to maintain Britain's place in the sun.

In the last two years of the war and in view of American aloofness from Europe Churchill became increasingly uneasy about Stalin's territorial ambitions, some of which might well be at the expense of the British Empire. Already in October 1942 he had warned his Foreign Secretary that it would be a

> "*measureless disaster if Russian barbarism overlaid the culture and independence of the ancient states of Europe*"[93].

A year later Churchill returned to his idea, first ventilated after the Russian revolution in 1917, of building up Germany as a buffer state between the Soviet Union and continental Europe. He explained to the British cabinet that we

> "*mustn't weaken Germany too much – we may need her against Russia*"[94].

This was reminiscent of the opinion he had expressed as early as December 1940, when he explained that one had to differentiate between the Nazi regime and the German people. He very much hoped that the victorious powers would succeed in ensuring that

> "*Germany was going to remain in the European family... there should be no Pariahs*". After all, "*Germany existed before the Gestapo*"[95].

---

92 *W. S. Churchill*, The World Crisis, Vol.1: 1911–14, London 1923, 181.

93 NAUK: PREM 4/100/7, PM's Personal Minute M.474/2 to Eden, 21/10/1942; also in *Churchill*, The Hinge of Fate, 504.

94 Quote: *T. H. Anderson*, The United States, Great Britain and the Cold War, 1944–47, Columbia, MO 1981, 10. On 23 Nov. 1954 Churchill gave a speech in Woodford outside London during which he claimed that at the end of the Second World War he asked Field Marshall Montgomery to make preparations for rearming German soldiers to use them against a further western advantage by the Red Army. The speech aroused much controversy and a few days later Churchill admitted that he had been mistaken and had never given such an order.

95 Quoted in *J. Colville*, The Fringes of Power: 10 Downing Street Diaries, 1939–55, London 1985, 312–13: diary entry, 13 Dec. 1940; see also *A. Montague Browne*, Long

In November 1943, on the way to the Tehran conference, he explained to Harold Macmillan that

> *"Germany is finished ... The real problem now is Russia. I can't get the Americans to see it"*[96].

By July 1944 the British Chiefs of Staff firmly supported him in this view. Ever since the German defeat at Stalingrad in early 1943 the COS had begun to express the opinion that the maintenance of amicable Anglo-Soviet relations rather than the remote likelihood of renewed German aggression would be crucial for the post-war period. The British military were increasingly suspicious of Stalin's intentions. Soon, their anti-Russian views became widely known and, at this stage, clashed with the somewhat more optimistic Foreign Office[97]. Churchill, however, considered the COS's scepticism fully justified. In August 1944 Lord Moran, his personal physician, jotted in his diary:

> *"Winston never talks of Hitler these days; he is always harping on the dangers of Communism. He dreams of the Red Army spreading like a cancer from one country to another. It has become an obsession, and he seems to think of little else."*[98]

However, as happened quite frequently, Moran was too impressed by one of Churchill's dramatic outbursts. Even in 1944–45 Churchill tended to be torn between deep mistrust about the territorial strategy of the Soviet dictator in central Europe and other regions and wishful thinking about Stalin's personality and the possibility of friendly co-operation with Moscow in the post-war years[99]. This would remain the case in the years from the end of the war until Stalin's death in early 1953. In January 1944, for example, after having recovered from his first heart attack, Churchill talked to Foreign Secretary Eden about "*the new confidence which has grown in our hearts towards*

Sunset: Memoirs of Winston Churchill's Last Private Secretary, London 1995, 159. See also *Gilbert*, Churchill, Vol. 6: Finest Hour, 943.

96 Quote: *Anderson*, The United States, Great Britain, 10–11. See also *J. Wheeler-Bennett and A. Nicholls*, The Semblance of Peace: the Political Settlement after the Second World War, London 1972, 290.

97 For the COS, see *L. Woodward*, British Foreign Policy in the Second World War, Vol. 5, London 1976, 189–90; *Ross*, Foreign Office Attitudes, 528–33; also *Erickson*, Stalin, Soviet Strategy and the Grand Alliance, 139.

98 *Lord Moran*, Churchill: The Struggle for Survival, 1940–65, London 1966, 173 (21/8/1944).

99 See *Barker*, Churchill and Eden, 221–32, 286–95. These conflicting views were also held by Foreign Secretary Eden who, in this regard, tended to be even more optimistic than his superior.

Stalin"[100]. Yet, in the spring of 1944 he became deeply concerned about the Red Army's advance into the Baltic states and northern Rumania and Poland to which he had consented at Tehran. The Prime Minister increasingly feared not only the 'communizing' of these countries and the whole of Eastern Europe but also the possibility, if not the likelihood, of Stalin intending to conquer the Balkans and the entire Mediterranean including Greece, Italy and perhaps even France. Only together with the USA would Britain be able to resist Stalin's territorial and ideological drive[101]. Churchill's confidence in Stalin was also temporarily shattered by the Warsaw Uprising in August and September 1944. The Prime Minister was horrified by Stalin's refusal to order the Soviet troops outside the city to intervene and thus prevent the wholesale slaughter of 200,000 Polish civilians by the German army and he was also shocked by Roosevelt's coolness regarding the whole terrible situation[102].

However, by late 1944, after the so-called 'Tolstoy' meetings with Stalin in Moscow, during which he entered into the notorious percentage deal with the dictator to contain Moscow's expansionist ambitions by means of an Anglo-Soviet pact to divide up South-eastern Europe, Churchill informed Roosevelt of the "*extraordinary atmosphere of goodwill*" he had encountered in Moscow. To Eden he spoke of Stalin as a "*great and good man*"[103]. And the Foreign Secretary agreed with this assessment. In conversation with the by then much more pessimistic Czechoslovakian President Benes Eden explained that "*despite occasional irritants*", on the whole he was convinced that "*the Russians genuinely wished to work*" with Britain[104]. After the Yalta Conference in February 1945, during which amongst many other issues de-

---

100 Quoted in *Douglas*, From War to Cold War, 11–12 (quote: 12). See also *Barker*, ibid., 223.

101 For Stalin's territorial conquests towards the end of the war, see for example in great detail *Weinberg*, World at Arms, 587 ff.; *Overy*, Russia's War, 237 ff.

102 See *Edmonds*, The Big Three, 384–86. However, later Churchill himself did not shrink from ordering the bombardment of German cities, as in the case of Dresden which caused the unnecessary death of many thousands of largely innocent civilians. Also his thinking about the use of poison gas (above all mustard gas) in the bombardment of German cities is well known. See *G. W. Gellermann*, Der Krieg der nicht stattfand (with the facsimile of a memo by Churchill on the issue, dated 6/7/1944), o. O. 1986, 249–51.

103 For a good comprehensive overview see *Churchill & Roosevelt* (note 63), 348–51. Quotes: ibid., 353; for the percentage deal with which the two politicians agreed on dividing up the Balkans between them, see *W. S. Churchill*, Second World War, Vol. 6, 197–212; *Edmonds*, Churchill and Stalin, 320–21; *Kimball* (note 22 – The Juggler), 159–83; *A. Resis*, The Churchill-Stalin Percentages Agreement, American Historical Review 82/2 (1978), 368–87.

104 NAUK: FO 371/43 336/N 7650/G, 27/11/1944.

cisions regarding the war against Japan as well as the four-power occupation of Germany, the new Russo-Polish border and the westward shift of Poland at the expense of Germany and the composition of the new Polish government had been taken, Churchill was still optimistic. He informed the British people that he knew

> *"of no Government which stands to its obligations… more solidly than the Russian Soviet Government".*

Churchill was convinced

> *"that Marshal Stalin and the Soviet leaders wish to live in honourable friendship and equality with the Western democracies"*[105].

On 17 July 1945, during the Potsdam Conference, the Prime Minister told Eden "*I like that man*"[106].

Despite Churchill's increasing geopolitical and ideological fears regarding Soviet intentions, the assumption appears to be correct that as late as the spring and summer of 1945, Churchill like Roosevelt and his successor Harry Truman

> *"still had some hope of friendship with Stalin, even if it was a badly battered hope"*[107].

Indicative of Churchill's attitude and duality of view in the closing months of the war was the message the Prime Minister sent Stalin in late April 1945 regarding Moscow's apparent intention to impose a communist government on Poland, annex the Baltic states, extend the Kremlin's dominance across most of central Europe and ignore altogether the Yalta agreements including his promise to hold free elections in Poland. Churchill asked Stalin:

---

105 Quoted in *Churchill*, Second World War, Vol. 6, 351. On 5 March 1945 Churchill wrote to Eden with respect to the strong criticism of the Foreign Office regarding Soviet activities in Roumania: 'We really have no justification for intervening in this extraordinary vigorous manner for our late Roumanian enemies ...'. Quoted in *D. Carlton*, Anthony Eden: A Biography, London 1981, 254. See also Kimball (ed.), ibid., 545–46. Eden appears to have been much more sceptical about Stalin at Yalta. See *R. R. James*, Anthony Eden, London 1987, 288–92. For the Yalta conference, see still *D. S. Clemens*, Yalta, New York 1970, and now *Fraser J. Harbutt*, Yalta 1945: Europe and America at the Crossroads, Cambridge 2010.

106 Quote: *R. M. Hathaway*, Ambiguous Partnership: Britain and America, 1944–47, New York 1981, 39. See also *Moran*, Struggle for Survival, 272 ff.

107 *Barker*, Churchill and Eden, 294; *Kimball*, Wheel within a Wheel, 302–05; see also *R. H. Ferrell*, The Dying President: Franklin D. Roosevelt, 1944–45, Columbia, Mo. 1998.

"But do not, I beg you, my friend Stalin, underrate the divergences which are opening about matters which you may think are small to us, but which are symbolic of the way the English-speaking democracies look at life."[108]

However, Stalin was not overly impressed.

Neither in the spring and summer of 1945 nor throughout the entire Second World War had Churchill been willing to accept that Britain and Europe might become subservient to Moscow in the post-war years. He was also strongly opposed to Britain being put in an inferior position to the United States for any length of time after the war. His position was, as he wrote Roosevelt in March 1944,

"that Great Britain seeks no advantage, territorial or otherwise, as the result of the war. On the other hand she will not be deprived of anything which rightly belongs to her after having given her best services to the good cause."[109]

While the Prime Minister greatly resented American anti-Colonialism and Washington's insistence on free trade and the removal of all post-war tariff barriers including the abolition of Britain's imperial preference system which would effectively lead to the dismantling of the Sterling area, he came to fear even more Stalin's apparent territorial ambitions in Eastern and Central Europe. Thus, instead of accepting any future subservience to either Washington or Moscow Churchill intended to maintain the independence of Britain and the European continent in its backyard and restore, as he saw it, Europe's strong and beneficial role in world politics.

Concluding Remarks

Despite some major strategic blunders, on the whole Churchill's performance as war leader was by no means as poor as is suggested in some of the recent literature. In fact, some of these strategic decisions and in particular his great emphasis on personal diplomacy and the development of constructive and cooperative relations with both the United States and the Soviet Union were decisive for the allied victory in the Second World War and certainly for Britain's part in that victory. It is doubtful that a differ-

108 Quoted in *Churchill*, Second World War, Vol.6, 433–34.

109 *Churchill & Roosevelt* (note 63), 17: Churchill to Roosevelt, 4/3/1944. There was a strong nationalistic streak in Churchill's imperialism; he was fully aware that without the Empire Britain would never have become a global power.

ent war leader would have obtained similar results for British politics and strategy during World War II.

Autorenverzeichnis/List of Contributors

Professor Derek Beales FBA, Sidney Sussex College, Cambridge, CB2 3HU, United Kingdom

Professor T. C. W. Blanning FBA, Sidney Sussex College, Cambridge, CB2 3HU, United Kingdom

Professor Christopher Clark FBA, St Catharine's College, Cambridge, CB2 1RC, United Kingdom

Professor Frank-Lothar Kroll, Technische Universität Chemnitz, Reichenhainer Str. 39, D-09126 Chemnitz, Germany

Professor Klaus Larres, University of Ulster, Cromme Road, Londonderry, BT52 1SA, Northern Ireland, United Kingdom

Professor Keith Robbins, Rhydyfran, Cribyn, Lampeter, Ceredigion SA48 7NH, Wales, United Kingdom

Professor Brendan Simms, Peterhouse, Cambridge, CB2 1RD, United Kingdom

Professor Thomas Stamm-Kuhlmann, Universität Greifswald, Historisches Institut, Allgemeine Geschichte der Neuesten Zeit, Bahnhofstr. 51, D-17487 Greifswald, Germany

PD Dr. Karina Urbach, Institute of Historical Research, Senate House, Malet Street, London, WC1E 7HU, United Kingdom

Professor Dieter Weiß, Präsident der Prinz-Albert-Gesellschaft, Universität Bayreuth, D-95440 Bayreuth, Germany